Pests, Diseases, and Disorders of Peas and Beans

A Colour Handbook

Anthony J. Biddle

Processors and Growers Research Organisation, Peterborough, UK

Nigel D. Cattlin

Holt Studios, UK

With a Foreword from
Dr John M. Kraft

formerly of USDA Agricultural Research Service, Washington, USA

CRC Press
Taylor & Francis Group
Boca Raton London New York

CRC Press is an imprint of the
Taylor & Francis Group, an **Informa** business

CRC Press
Taylor & Francis Group
6000 Broken Sound Parkway NW, Suite 300
Boca Raton, FL 33487-2742

© 2007 by Taylor & Francis Group, LLC
CRC Press is an imprint of Taylor & Francis Group, an Informa business

First issued in paperback 2019

No claim to original U.S. Government works

ISBN 13: 978-0-367-45314-5 (pbk)
ISBN 13: 978-1-84076-018-7 (hbk)

**Visit the Taylor & Francis Web site at
http://www.taylorandfrancis.com**

**and the CRC Press Web site at
http://www.crcpress.com**

A CIP catalogue record for this book is available from the British Library.

ISBN-13: 978-1-84076-018-7

Plant Protection Handbooks Series
Alford: *Pests of Fruit Crops – A Colour Handbook*
Alford: *Pests of Ornamental Trees, Shrubs and Flowers – A Colour Atlas* Biddle & Cattlin: *Pests and Diseases of Peas and Beans – A Colour Handbook* Blancard: *Cucurbit Diseases – A Colour Atlas*
Blancard: *Tomato Diseases – A Colour Atlas*
Blancard: *Diseases of Lettuce and Related Salad Crops – A Colour Atlas* Fletcher & Gaze: *Mushroom Pest and Disease Control – A Colour Handbook* Helyer *et al*: *Biological Control in Plant Protection – A Colour Handbook*
Koike *et al*: *Vegetable Diseases – A Colour Handbook*
Murray *et al*: *Diseases of Small Grain Cereal Crops – A Colour Handbook* Wale *et al*: *Pests & Diseases of Potatoes – A Colour Handbook*
Williams: *Weed Seedlings – A Colour Atlas*

Contents

Foreword 5

Preface 6

Acknowledgements 6

Glossary 7

Section 1 Introduction 9
Peas and beans in agriculture 10
Pea and bean production 12
Quick guide to diagnosis 16

Section 2 Diseases of Seedlings and Young Plants 19
Pythium seedling rot: *Pythium* spp. 20
Pythium and *Rhizoctonia* spp. root rot: *Pythium ultimum* and others, and *Thanatephorus cucumeris* syn. *Rhizoctonia solani* 22
Black root rot: *Thielaviopsis basicola* 23

Section 3 Pests of Seedlings and Young Plants 25
Wireworms: *Agriotes* spp. 26
Bean seed fly (seed corn maggot): *Delia platura* 27
Pea and bean weevil: *Sitona lineatus* 28
Field thrips: *Thrips angusticeps* 30
Leatherjackets: *Tipula* spp. 32

Section 4 Fungal and Bacterial Diseases 35
Alternaria leaf spot: *Alternaria alternata* 36
Common root rot: *Aphanomyces euteiches* 37
Bean leaf and pod spot: *Ascochyta fabae* = *Didymella fabae* 38
Pea leaf and pod spot: *Ascochyta pisi, Mycosphaerella pinodes*, and *Phoma medicaginis* var. *pinodella* 40
Botrytis pod rot: *Botrytis cinerea* 42
Chocolate spot: *Botrytis fabae, B. cinerea* 44
Anthracnose: *Colletotrichum lindemuthianum* 46
Powdery mildew: *Erysiphe pisi* 48
Fusarium wilt: *Fusarium oxysporum* f. sp. *pisi* Races 1, 2 and 5 50

Pea foot rot: *Fusarium solani* f. sp. *pisi* and *Phoma medicaginis* var. *pinodella* 52
Fusarium root rot (Fusarium yellows): *Fusarium solani* f. sp. *phaseoli* 54
Bean foot rot: *Fusarium solani, Phoma medicaginis* var. *pinodella*, and *Fusarium culmorum* 55
Bean downy mildew: *Peronspora viciae* 57
Pea downy mildew: *Peronspora viciae* 59
Ascochyta leaf spot: *Phoma exigua* var. *exigua* = *Ascochyta phaseolorum, Ascochyta bolthauseri* 61
Halo blight: *Pseudomonas syringae* pv. *phaseolicola* 62
Pea bacterial blight: *Pseudomonas syringae* pv. *pisi* 64
White mould, Sclerotinia: *Sclerotinia sclerotiorum* 66
Stem rot: *Sclerotinia trifoliorum* 68
Bean rust: *Uromyces appendiculatus* 69
Bean rust: *Uromyces fabae* 70
Common blight: *Xanthomonas campestris* pv. *phaseoli* 72

Section 5 Viral Diseases 73
Alfalfa mosaic virus (AMV) 74
Broad bean stain virus (BBSV) and broad bean true mosaic virus (BBTMV) 75
Bean common mosaic virus (BCMV) 76
Bean curly top virus (BCTV) 77
Bean leaf roll virus (BLRV) 78
Bean yellow mosaic virus (BYMV) 79
Cucumber mosaic virus (CMV) 80
Pea early browning virus (PEBV) 81
Pea enation mosaic virus (PEMV) 82
Pea seed-borne mosaic virus (PSbMV) 84
Pea streak virus (alfalfa mosaic/red clover vein virus) (PSV) 86
Pea top yellows virus (PTYV) syn. pea leaf roll virus, bean leaf roll virus (BLRV) 87

Section 6 Pests of Stem, Foliage, and Produce   89
Pea aphid: *Acyrthosiphon pisum*   90
Cutworms: *Agrotis segetum*   92
Black bean aphid: *Aphis fabae*   93
Silver Y moth: *Autographa gamma*   95
Pea seed beetle: *Bruchus pisorum*   96
Bean seed beetle: *Bruchus rufimanus*   98
Tortrix moth: *Cnephasia asseclana*   100
Pea midge or pea gall midge: *Contarinia pisi*   101
Pea moth: *Cydia nigricana*   103
Slugs and snails: *Deroceras reticulatum,*
   Cernuella sp.   105
Stem nematode: *Ditylenchus dipsaci*   107
Heliothis caterpillar (corn earworms):
   Helicoverpa armigera and other
   Helicoverpa spp.   108
Pea cyst nematode: *Heterodera gottingiana*   110
Pea thrips: *Kakothrips pisivorus*   112
Leaf miners: *Lyriomyza* spp.   113
Root knot nematode: *Meloidogyne* spp.   114
Two-spotted spider mite: *Tetranychus urticae*   115
Stubby root nematode: *Trichodorus* spp. and
   Paratrichodorus spp.   116

Section 7 Seedling and Crop Disorders   119
Hollow heart   120
Manganese deficiency   120
Seed vigour   122
Sulphur deficiency   123
Water congestion   124
Iron deficiency   124

Further Reading   126

Index   127

Foreword

It has been my pleasure to have known Anthony Biddle both professionally and personally for over 20 years. I have been impressed with his thorough knowledge and dedication in working with the pea and bean industry in the UK, Europe, and North America. I have also been impressed with the respect that this industry has shown him over the years. This handbook is a logical extension of that dedication and hard work. It is a thorough compilation of information on the diagnosis and control of pests, diseases, and disorders of peas, faba beans, and common beans in one fully illustrated, easy to read and understand publication. The handbook is written in a clear and concise style accompanied with relevant high quality photographs of these problems which allows the reader to assess and diagnose problems quickly on these crops. Nigel Cattlin has produced illustrations of all the main crop problems and their clarity and detail will greatly assist these diagnoses. The disease or disorder is listed along with the host crop, symptoms, economic importance, disease cycle, and control. The sections are listed in a logical order starting with seedlings and young plants and ending with seed and crop disorders. The reader can refer quickly to the appropriate section to seek an answer to a particular production problem. The Quick Guide to diagnosis is a valuable and concise tool for a production specialist to diagnosis a field problem. The section on insect pests of seedling and young plants condenses a voluminous amount of literature and data into a readable and clearly understandable section. The section on fungal and bacterial diseases covers all the minor and major diseases of peas, faba bean, and common bean on a worldwide basis. The sections on viral diseases, pests of foliage and produce, and seedling and crop disorders are thorough and concise. This handbook is a must-have publication for all advisory plant specialists, growers, seedsmen, production specialists, diagnostic clinicians, and agribusiness representatives who have an interest in these crops. I highly recommend it.

John M. Kraft

Preface

Large seeded legumes, cultivated as pea and beans, are an important crop worldwide, both as high value vegetables or as a dry harvest crop which is high in protein and a staple addition to the diet. Peas and beans are also important crops in a farming rotation as they allow a break from cereal production in arable areas and the residual nitrogen produced by peas and faba beans are utilized effectively by any following crop. Intensive production can often result in an increase in pest or disease pressure thereby putting crops at risk either from yield losses or reduction of quality. Such problems are often resolved by the use of pesticides or of cultural methods of control or avoidance. It is important that problems are recognized early on and correctly identified and assessed as a possible risk before any treatment is applied. Identification of these problems is therefore a key part of crop protection and crop management. However, there is a shortage of experienced field pathologists and entomologists and those who are engaged in the identification and advice to growers are less crop-specialized. Assistance with identifi-cation in the field is therefore a welcome aid to all who are engaged in this work and specialist publications covering specific crop problems are considered key assets.

I have specialized in the research, identification, and control of pests and diseases of the pea and bean crops for more than 30 years. During this time I have been in contact with many growers and advisers and decided that a distillation of this experience would form an integral part of a pest and disease book for legume crop protection, with descriptions of symptoms linked to detailed illustrations and notes on prevention and control. Nigel Cattlin has an international reputation for the production of high quality photographs and his illustrations are an essential part of this book. The approach has been to cover a wide range of common and less common problems in a book that will be easy to use as a field guide for growers, advisers, and extension workers as well being as useful reference for researchers.

Anthony J. Biddle

Acknowledgements

Anthony Biddle is indebted to the help given by colleagues at Processors and Growers Research Organisation (PGRO), particularly Becky Ward, and earlier by Barry McKeown for help in dealing with the many enquiries and investigations concerning pea and bean problems over the years. Also to Martin Whalley of Manchester Metropolitan University for his help and identification of several of the fungal pathogens, and Richard Larsen and Phil Miklas of USDA Centre, Prosser, Washington State, USA for information and illustrations of bean diseases. Thanks also to John Kraft, formerly of USDA, Prosser, for advice and hospitality when the author was able to visit crops and growers over several seasons. To John Kraft and Jane Thomas for commenting on the manuscript, and to all UK growers of peas and beans who have allowed the author to record their crop problems. Finally, to the support that PGRO has given the author to produce this book.

In addition to adding my thanks to everyone at PGRO for their help over many years, Nigel Cattlin would also like to extend his thanks to Professor Antonio Monteiro and Mrs Luisa Moura for their invaluable help with photography for this book in Portugal.

Glossary

abscise leaf fall

adaxial the flat internal surfaces of the hemispherical portions of a legume seed

adventitious developing from the shoot or root

afila a leaflet converted to a tendril

alate with wings

antenna(e) the sensory jointed appendage to the head of an insect

aphicide an insecticide for controlling aphids

apical shoot the terminal or youngest upper shoot

apterous without wings

biological control the control of a pest or disease using natural methods

caterpillar the larva of a moth or butterfly

chlamydospore an asexually-produced thick-walled resting spore

chlorosis yellowing

chrysalis the hard case enclosing a pupating caterpillar

cocoon the case enclosing a pupating insect larva

colonize the spread of an organism on its host away from the initial site of infection or infestation and the dependence on the host for nutrients

compaction the consolidation of soil

conidia asexually-produced fungal spores

conidiophore a specialized hyphal branch bearing conidia

cotyledon a seed leaf, the two hemispherical portions of a legume seed

cuticle the superficial skin or outer layer

cyst the hardened outer case containing eggs or larvae of nematodes

damping-off the rot of seedlings near soil level after or before emergence

desiccate to dry up

diurnal active during the day

enation a small protruding growth

epidemic a progressive increase in the incidence of a particular disease

epidermis the outer layer of tissue

exudate a substance passed from within a plant or seed to the outer surface

herbicide a pesticide used for controlling weeds

honeydew the sticky substances excreted by aphids during feeding

host an organism harbouring a parasite

hypha(e) the tubular thread-like filament(s) of fungal mycelium

hypocotyl the region between the root and the stem

imbibe the absorption of moisture by a seed

inoculum spores or other pathogen parts which can initiate disease

instar the specific development stages of an insect larva

integrated crop management the use of good husbandry techniques which include justifiable use of pesticides for crop production

internode the area of stem between nodes

interveinal between the leaf veins

larva(e) the immature stage(s) of an insect

leachate the liquid which has filtered out

leaf axil the angle between the leaf and the stem

lesion a localized area of diseased or disordered tissue

micropyle a minute pore in the seed coat

mosaic the patchy variation of normal green colour symptomatic of many viruses

molluscicide a pesticide for the control of molluscs (slugs and snails)

moribund at the point of death

mycelium a mass of hyphae that form the vegetative part of a fungus

necrotic the browning or blackening of cells as they die

node the part of the stem to which the leaf is attached

nodule a growth on the root formed by *Rhizobium* bacteria

nymph an immature insect similar to the adult in appearance

oospore the asexually-produced resting spore of fungi in the class of oomycetes

pathogen an organism which causes disease

peduncle the stalk of a flower or pod

perfect state the sexual reproductive stage in the life cycle of a fungus

persistent virus a virus which persists in a vector for >100 hours, and in some cases for the life of the vector

pH a measure of acidity or alkalinity

pheromone a chemical produced by an insect as a means of communication

plumule an embryonic shoot

pulse the seed of large-seeded legumes

pupa the resting stage of insect metamorphosis

pustule a blister-like spore mass breaking through the plant epidermis

pycnidia flask-shaped or spherical receptacles bearing asexual spores

pycnospore an asexual spore produced in a pycnidium

race a strain of fungal or bacterial pathogen that is morphologically or physiologically indistinguishable, but pathogenically specialized to different varieties of host species

resistant variety a variety which has the ability to prevent or retard the development of a given pathogen or pest

saprophytic mould a fungus living on dead and decaying material

sclerotium a long-lived, compacted mass of vegetatively-produced hyphae

senescence ageing which eventually leads to death

spore a specialized propagative or reproductive body in a fungus

stamen sheath the covering of the male sexual parts of a flower

stipule a leaf-like outgrowth from the base of a true leaf

systemic pesticide a pesticide which is absorbed and translocated in the plant

systemic infection an infection that spreads throughout the plant from a single infection point

testa the seedcoat

trifoliate the first true leaf of a *Phaseolus* species (three leaflets)

vascular strands the liquid-conveying vessels or tubes within a stem or root

vector an organism which transmits a pathogen, usually a virus

viviparous the hatching and development of larvae within the parent insect

zoospore a fungal spore capable of movement in water

SECTION 1

Introduction

- PEAS AND BEANS IN AGRICULTURE

- PEA AND BEAN PRODUCTION

- QUICK GUIDE TO DIAGNOSIS

PEAS AND BEANS IN AGRICULTURE

Peas and beans as fresh and dry pulses have been in cultivation for as long as man has been able to cultivate the land. Peas and beans are legumes; the species belong to the family Fabacae, of which there are many members worldwide. They have been used as food for man, either in the fresh or green state for centuries. They were also one of the first 'convenience foods' since, when dried naturally in the field as if for seed, they store well and provide a nutritious and filling vegetable (1). This book describes the most important pests, diseases, and disorders of this large, seeded, legume crop.

Peas probably originated in Middle Asia and the central plateau of Ethiopia (2). By the Bronze Age, they were used by inhabitants of central Europe and primitive seeds have been found in areas inhabited by Swiss lake dwellers and in caves in central Hungary. Peas were known by the Greeks and Romans and were first mentioned in England after the Norman Conquest (1066 AD). Peas are widely grown in Europe, both as a fresh vegetable and as a dried pulse crop.

Beans too have their origins deep in history. There were frequent references to beans (3) by the ancient Egyptians, Greeks, and Romans and remains of an early crop were excavated at an iron age site in Glastonbury, England. *Vicia faba* beans are still grown in the Mediterranean region despite a proportion of the population having an allergy to the crop known as 'favism'. It is thought that Pythagoras met his death in the sixth century BC through being caught by his enemies because of his inability to cross a bean field.

Vicia faba is an important crop in China, western Europe, Australia, and the Middle East, the latter country using dried beans as an important part of the daily meal. Again, this species has been exploited to provide a range of varieties for use both as a fresh vegetable known as broad beans, or for types harvested dry as seed. In Europe, the dried Vicia beans and dried peas are a useful source of protein which can be included in animal feeding rations.

Phaseolus beans (4) are thought to have originated from South America and are in the main still favoured by warm growing conditions. The

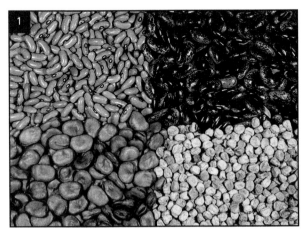

1 Peas and bean seeds.

2 Peas (*Pisum sativum*) at pod fill.

beans are grown either as dry harvested crops for seed, or harvested fresh where the pod is eaten whole. Many different forms are grown, either as a bush type or a climbing bean with intermediate types grown for specific produce requirements.

Species include *Phaseolus vulgaris*, the dwarf french, dwarf green, bush, or snap bean and *P. lunatus*, the lima bean. A further species, *P. coccineus*, is grown as a climbing type, often being supported on sticks or wires, and is known as the runner bean (5).

3 Field bean (*Vicia faba*) pods.

4 Phaseolus beans (*Phaseolus vulgaris*).

5 Runner beans (*Phaseolus coccineus*) flowers and beans.

PEA AND BEAN PRODUCTION

Modern types of peas and beans are grown for both human consumption and animal feed, where they are relatively high in protein and used in animal rations to supplement soya or other protein. Peas grown for dry harvest are known in Europe as combining peas. Modern varieties are mostly afila (semi-leafless) types, where the leaflets have been modified to tendrils providing an improved standing ability and suitability for mechanical harvesting. The varieties produce seed which is either smooth or dimpled and all types have their own specialized market niches. Peas harvested fresh are known as vining peas, where the pods are mechanically threshed green in the field and the peas are processed as frozen or canned peas within a few hours of harvest. Those harvested green for the pods are known as garden or fresh market peas and are usually harvested by hand. An additional small quantity is grown for whole pod consumption and are known as mange-toutes, sugar snaps, or snow peas. Since the development of canning and freezing techniques, peas are considered an important vegetable for use on their own or in mixed vegetable packs or in 'ready-meals'.

In the USA, peas were first brought over by European immigrants in the 1500s. From New England, settlers moving west by wagon train brought the crop first to Wisconsin where the canning industry began, to the west where they are now grown in the states of Idaho, Washington, and Oregon as dried and green peas (6). The first steps to conserving fresh peas came in 1885 at the Paris Exhibition, when Madam Faure exhibited a hand-operated pea viner, which shelled fresh peas. The basic principles of this machine are still used in modern day self-propelled field viners (7). The next great advance was the development of preservation by canning. The vegetable canning industry started around the turn of the twentieth century.

Frozen peas first began to be processed in the early 1930s primarily by fish freezers, hence the development of freezing factories in major fishing ports. However, the major growth in frozen peas was in the 1950 to the mid-1970s, with the increase in consumption linked to demands for convenience foods and the ownership of home freezers. The amount of peas grown for canning has now declined since the early 1930s and, today, around 65% of vining peas are destined for quick freezing.

Pea varieties have been improved by careful breeding. The first peas were pale in colour and susceptible to poor weather conditions and disease.

6 Centre pivot irrigation of peas in the Columbia basin, USA. (Courtesy of PGRO.)

7 Harvesting vining pea crop. (Courtesy of PGRO.)

Modern varieties are now very tolerant to disease, are high yielding, with generally improved flavour. A range of types are available to provide a 6–8 week harvesting period, allowing a continuity of fresh product to the factory for freezing or canning.

Harvesting has now become largely mechanized. By the 1950s, Dutch barns with factory owned static viners were seen on many farms in Lincolnshire and East Anglia. The use of such machines was based on the peas being cut in the field and loaded onto a trailer and transported to the viner. The next step was to introduce a trailed viner which picked up a cut crop in the field and vined the peas into a trailer. Since then complete harvesters, first seen in the 1970s, are now the main form of harvesting, with vined peas being delivered to the factory within an hour of harvesting. Control of harvesting is now largely based on the accurate measurement of maturity. In the early 1950s, the Martin Pea Tenderometer was invented which even today, provides the farmer and factory with a robust means of assessing the optimum time for vining.

The main bean types are *Vicia faba* and *Phaseolus* spp. *Vicia faba* or field beans are also grown as dry harvested crop, mainly for inclusion as a protein ration in animal feed, but a proportion are sold for human consumption used in the Middle East. Where *Vicia faba* varieties have been bred for harvesting fresh as a vegetable, these are known as broad beans. The vined beans are either frozen or canned or marketed in the pods for the fresh vegetable market.

Phaseolus beans are mainly *P. vulgaris* types. In Europe, they are harvested green for their pods and are either frozen or canned. In southern Europe, the USA, and Canada a large crop is harvested dry as seed, and is used for reprocessing in soups, salads, or canned in tomato sauce as 'baked beans'. A significant proportion of dry harvested types is grown in southern Europe and South America for hand harvesting and is used locally. *P. coccineus,* the runner bean, has a specialist fresh market in Europe but is grown for hand harvesting in the UK, Africa, and India.

Peas and beans are grown mainly in arable rotations, although the specialist markets for fresh produce are confined to relatively intensive vegetable production areas. Peas and Vicia beans are favoured as break crops in mainly cereal rotations. The residual nitrogen fixed by the *Rhizobium* bacteria in the root nodules (8) benefits the following autumn sown cereal crop. The ability to plant the peas or beans in the spring (9) allows land to be fallow for the winter and provides opportunities to control perennial weeds and grass weed species which is difficult in a cereal rotation. A proportion of Vicia

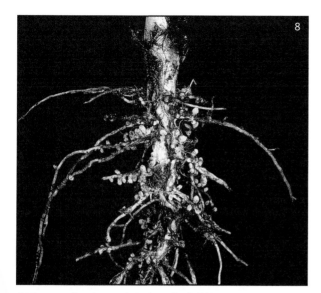

8 Nitrogen fixation nodulation of a field bean root.

9 Drilling pea crop. (Courtesy of PGRO.)

beans are planted in the autumn, especially in the UK, where they survive as seedlings until the spring when growth is rapid and offers the chance of a relatively early harvest of mature beans (**10**).

The nutritional requirements of peas and Vicia beans are small. Nitrogen is unnecessary and the amounts of potassium, phosphate, and sulphur are also small compared with more nutrient demanding crops such as brassicas, potatoes, or sugar beet. Peas on high alkaline soils can become deficient in manganese. The recent lessening deposition of sulphur is resulting in sulphur deficient crops on lighter free-draining soils. Phaseolus beans, in Europe at least, have a requirement for nitrate as the naturally occurring *Rhizobium* bacteria are not present in many soils. In the USA, soils are often deficient in zinc and molybdenum.

Crop Protection

All crops are sensitive to soil compaction, pan, smear, capping, and waterlogging. Poor soil structure results in low populations of plants which are incapable of reaching their full development and which are susceptible to root diseases. Short rotations encourage the development of soil-borne diseases and pests, which in turn depress yields. Rotational practice should be planned carefully and peas and beans should be considered as one and the same crop and not grown in a field any more frequently than once in 5 years.

All types require a well-drained seed bed for rapid germination and establishment (**11**). However, even in the best of situations efficient and profitable crop production is dependent upon many factors. In particular, many pests, diseases, and disorders are capable of reducing crop yields, spoiling quality, jeopardizing the reliability of production on the farm, and disrupting throughput at the factory. In crops for human consumption, crop treatment may not be justifiable for yield increases alone, but product quality and contamination are leading reasons for control measures to be carried out.

The likelihood of pest, disease, or disorder should always be considered when cropping is planned. It should be considered in relation to soil type, rotation, and problems in previous crops. Very occasionally there are dramatic losses due to infestations or infections of a particular pest or disease, which may often be associated with climatic conditions. There are less serious losses, such as when patches of diseased plants appear or when pest attack warrants treatment but is not particularly heavy. In all situations, it is the producer who needs to be vigilant and monitor the crops frequently throughout the growing season. This book describes the problems that may be encountered during the season, the symptoms and economic impact, and also describes means of control by both cultural and chemical means.

The use of crop protection chemicals is always under scrutiny. Without the use of these, production of high yielding, high quality crops is much more difficult. However, there is an increasing emphasis on a more sustainable means of production and that in turn leads to the more prudent use of agrochemicals for crop protection. The basic biology of pests and diseases have led to the improvement in monitoring, disease-resistant varieties, and rotational factors which will help to avoid serious problems. The use of all these techniques, coupled with good husbandry practices, will inevitably lead to less reliance on pesticides. The use of monitoring systems (**12**) and forecasting models checking for treatment thresholds before pesticide applications are justified are all necessary parts of integrated crop management (ICM)(**13**). The following chapters make use of these techniques to avoid problems where possible, but allow correct identification of the problem to be made and to justify treatment, if necessary.

10 Combining field beans.

11 Epigeal germination of *Phaseolus vulgaris* bean.

12 Pest monitoring with a pheromone trap.

13 Applying insecticide to very young pea crop.

QUICK GUIDE TO DIAGNOSIS

SYMPTOMS	POSSIBLE CAUSES
PEAS	
Seedlings	
Death of growing point, brown spot in the centre of the cotyledon	Marsh spot caused by manganese deficiency
Primary shoot with black girdling of adjoining seeds	Pea foot rot (Mycosphaerella/Phoma foot rot)
Weak seedling centre of cotyledons with deep cavities	Hollow heart
Rotting at or below soil surface	Damping-off (Pythium seedling rot)
Shoots severed just below soil surface, secondary shoots developing	Leatherjackets
Shoots and seeds with narrow holes and tunnels	Wireworm
Leaves torn, stems rasped	Slugs or snails
Deep sunken cavity in the centres of the cotyledons	Hollow heart
Young plants	
U-shaped leaf notches	Pea weevil
Yellowing crinkled leaves, translucent spots	Field thrips
Circular brown spots on leaves	Ascochyta leaf spot (pea leaf and odd spot)
Yellowing stunted plants, grey/mauve mould beneath leaves	Pea downy mildew
Water-soaked, slimy stems, later with white mould	White mould (*Sclerotinia* sp.)
Water-soaked leaf and pod spotting	Pea bacterial blight
Yellowing interveins and leaf margins	Manganese deficiency
Yellowing foliage	Sulphur deficiency/Iron deficiency
Purple-brown flecks and streaks	Pea leaf and pod spot (*Mycosphaerella* sp.)
Rotting of outer layer of root tissue	Common root rot (*Aphanomyces* sp.)
Stunted plants, root proliferation, and shortened growth	Stubby root nematode
Webbing of upper foliage	Tortrix moth caterpillar
Flowering plants	
Nettlehead, white maggots in buds	Pea midge
Plants grey-green, leaf roll, death in patches	Pea wilt (Fusarium wilt)
Foliage with white blisters or snaking tunnels	Leaf miner
Lower leaves yellow then death in patches, xylem reddening	Fusarium foot rot (pea foot rot)
As above, lower stem black	Pea foot rot (*Phoma* sp.)
As above, roots black	Black root rot
Stunted, stiff, yellowing plants, tiny white to brown cysts on roots	Pea cyst nematode
Browning and narrowing of leaf tips	Water congestion
Fine white film over foliage	Powdery mildew
Stipules narrow, mosaic leaves, translucent spotting, distorted pods	Pea enation, mosaic virus

Symptoms	*Possible causes*
Upper foliage yellow	Pea top yellow virus
Leaves pale and mottled, pods flat and purple-brown	Pea streak virus
Greenfly colonies in shoots	Pea aphid
Pale yellow plant, upper part more yellow than older leaves	Sulphur deficiency
Partial wilting, leaf marbling, stunting and death of growing point	Pea early browning virus

Podded plants

Shortened growth, downward leaf roll, small pods, blistered seeds	Pea seed-borne mosaic virus
Top leaves and pods curling, sticky exudate	Pea aphid
End of pods rotting, grey mould	Botrytis pod rot
Pods partially eaten by green caterpillar	Silver Y moth caterpillar
Pod surface blackened or silvery, roughened	Pea thrips

Produce

Holes in peas, creamy caterpillar with brown head	Pea moth
Circular holes in peas	Pea seed beetle
Brown spotting on seed surface	Pea leaf and pod spot (*Ascochyta* sp.)
Brown spot in cotyledon centre	Marsh spot
Blisters or 'tennis ball' markings on seed coat	Pea seed-borne mosaic virus

VICIA BEANS

Seedlings or young plants

U-shaped notches on leaf margins	Pea and bean weevil
Circular brown lesions with light grey centres	Bean leaf and pod spot (*Ascochyta* sp.)
Blackened stems at soil level, white mycelium	Stem rot (*Sclerotinia* sp.)
Small brown spots on leaves	Chocolate spot

Flowering plants

Pale blotches on leaf, velvety grey-mauve fungal growth beneath	Bean downy mildew
Small dark brown spots on leaves with larger, darker brown/grey lesions	Chocolate spot, *Cercospora*, *Stemphylium*
Dense colonies of black fly on upper stems	Black bean aphid
Brown twisted and swollen stems, distorted leaves	Stem nematodes
Black stems at soil level with pink fungal growth	Bean foot rot (Fusarium stem rot)
Pink larvae in diseased stems	*Reseliella* midge
Plants dying prematurely, blackened stem base	Bean foot rot (Fusarium or Phoma foot rot)
Interveinal yellowing of leaves, upper leaves rolled	Bean leaf roll virus
Fleshy white grubs feeding on root nodules	Pea and bean weevil larvae
Upper leaves with mottling or irregular-shaped dark green 'islands'	Bean yellow mosaic, broad bean stain, broad bean true mosaic, pea enation mosaic viruses

Mature plants

Orange-brown raised spots on leaves	Bean rust

SYMPTOMS	POSSIBLE CAUSES
Produce	
Deep circular brown/black spots on leaves, stems or pods	Bean leaf and pod spot (*Ascochyta* sp.)
Large circular holes in seeds, sometimes with windows in seed coat	Bean seed beetle
Darkened stain around edges of seed	Broad bean stain virus
PHASEOLUS BEANS	
Seedlings and young plants	
Severed shoots	Latherjackets, cutworms
Seedlings malformed with damage to growing point. White maggot feeding within seed or stem	Bean seed fly
Flowering and pod setting	
Reddish Discolouration and Shrivelling at Stem base	Rhizoctonia Root Rot
Deep lesions on stems and pods with brown centres containing orange spores	Anthracnose
Brown spotting of foliage, dark brown flecks on pods	*Ascochyta/Alternaria*
Leaves with irregular-shaped pale and dark green areas, leaves curl downwards. Brown discolouration of leaf veins	Bean common mosaic virus
Dwarfed plants with severe puckering and downward rolling of leaves. Plants dwarfed and bunched, leaves brittle, aborted flowers, pods stunted	Bean curly top virus
Domed areas on leaf surface, pods severely deformed and bent	Cucumber mosaic virus
Mosaic over leaf surface, chlorosis of foliage	Alfalfa mosaic virus
Grey mould on end of pods or around stems	Botrytis pod rot
Yellow patches on leaves, fine webbing underneath with small insects	Two-spotted spider mite
Small raised dark brown spots on leaves and pods	Phaseolus rust
Leaves with small brown irregular flecks surrounded by pale yellow edge	Halo blight
Leaf spotting with brown irregular lesions surrounded by bright yellow edge	Common blight
Patches of plants stunted with yellowing lower leaves	Fusarium foot rot (Fusarium yellows)
Roots decayed and blackened	Black root rot
Stunted plants, yellowing foliage, swollen areas on roots	Root knot nematode
Produce	
Seed with round holes or windows in seed coat	Bean seed beetle
Greasy spotting and some distortion of pods	Halo blight
Pods with deep holes or surface grazing	Silver Y, Heliothis caterpillars
Pods with red-brown flecks	Alternaria leaf spot
Leaves and pods with small brown circular lesions	Ascochyta leaf spot
Decayed areas on pods with grey mould	Botrytis pod rot
Pods with deep dark centred lesions containing orange spots	Anthracnose

SECTION 2

Diseases of Seedlings and Young Plants

- PYTHIUM SEEDLING ROT

- *PYTHIUM* AND *RHIZOCTONIA* SPP. ROOT ROT

- BLACK ROOT ROT

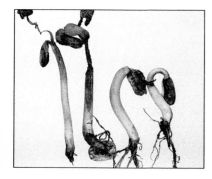
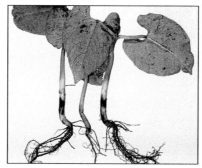

PYTHIUM SEEDLING ROT: *Pythium* spp.

Host Crops

Peas, green beans, runner beans, field beans, broad beans.

Symptoms of Infection

Peas that are sown early in the spring when soil temperatures are low and moisture is high, are usually slow to emerge. This is particularly the case with the wrinkled seeded vining or fresh market peas. Under these conditions, the seed is susceptible to infection by various soil-borne fungi including *Pythium* spp. *Pythium ultimum* is often associated with such infection. The cotyledons begin to decay shortly after the emergence of the radical. If germination occurs, then the young rootlets become infected and the seedling often fails to emerge from the soil (**14–16**).

The first sign of a problem is a poor seedling establishment, resulting in a low plant population (**16**). By digging along the rows, the remains of the seedcoat may be found together with a wet rotted cotyledon. Seedlings may have decayed hypocotyls and rootlets.

Economic Importance

A reduction in seedling population affects the growth and final yield of the crop. Poorly established crops allow an uncontrolled growth of weeds and are often more likely to suffer additional plant losses as a result of bird damage.

Disease cycle

Pythium spp. are attracted to exudates from imbibing seeds and quickly colonize a part of the cotyledon that is not protected by the seedcoat. Seed of low vigour leaches high levels of seed exudates during imbibition. Leachates containing carbohydrates and potassium salts are lost from damaged areas of the cotyledon and attract *Pythium* to the zone around the germinating seed. Once the fungi have colonized the tissue, they become aggressively pathogenic and cause pre-emergence seed rot or damping-off of emerged seedlings.

Prevention and Control

Seed from coloured flowered varieties contain anthocyanins and phenols which have some fungicidal effect. Such varieties are much less susceptible to seedling rot. White flowered varieties contain none of these compounds, although smooth seeded varieties are less susceptible as they contain less sugars to exude during imbibition.

Planting in warm, free-draining soils allows rapid germination and emergence of susceptible seed. Using seed of high vigour also reduces the risk of such losses (see Section 7). Most commercial varieties of white flowered varieties are treated with a fungicidal seed treatment which gives adequate protection from soil-borne damping-off diseases. Thiram, metalaxyl, or cymoxanil are all effective chemicals, some of which are commonly used in Europe, Australasia, and the USA for routine seed protection.

14 Damping off caused by *Pythium* spp.

15 Stem rot of Phaseolus beans caused by *Pythium* spp.

16 Pre-emergence pea seedling failure caused by
Pythium spp.

PYTHIUM AND RHIZOCTONIA SPP. ROOT ROT:
Pythium ultimum and others, and Thanatephorus cucumeris syn. Rhizoctonia solani

Host Crops
Green beans, dry beans, runner beans, peas.

Symptoms of Infection
Pythium is usually associated with pre-emergence seedling rot, and can affect the emergence and establishment of beans when sown in wet soil conditions. However, *Pythium* can also be found on the upper parts of young seedlings particularly during warm wet periods. The disease is known as tip blight, and the top part and growing points collapse following the development of a watery rot.

Rhizoctonia causes a stem rot of young seedlings, and is noticed after emergence when soil conditions may have been consolidated on the surface by heavy rain or excessive irrigation. The stem base is shrivelled with deeply sunken lesions which may completely girdle the stem (**17**). The lesion edges may be slightly red in colour and small black sclerotia may develop on the surface. Infected plants wilt and die.

Economic Importance
Both diseases are seldom of importance as they rarely cause widespread infection throughout the crop. However, a reduction in plant population overall will suppress the yield potential of the crop.

Disease cycle
Pythium spp. overwinters as oospores or saprophytic mycelium. Cool wet soil conditions favour infection of young roots. In warmer wetter conditions, rain splashed spores from infested soil can lodge in the upper leaf axils of peas or green beans, and when temperatures are around 18–28°C, tip blight is likely to occur.

Rhizoctonia overwinters as sclerotia or as chlamydospores. It can be spread from field to field by soil movement and by rain or irrigation water.

Root rot occurs during warm periods and is most severe at 15–18°C. Infection begins when the mycelium produced by the resting bodies infects the outer layer of the roots and invades the epidermis. Once infection has begun, the fungus spreads rapidly throughout the cells and tissues; lesions develop on the outside of the root before producing sunken areas containing small sclerotia.

Prevention and Control
Seed which is slow to germinate is more susceptible to *Pythium* infection, especially when soil conditions are wet and cold. Old seed may also be more susceptible to infection as the vigour of the seed may be reduced. Fungicidal seed treatment is normally made to processing varieties of beans, but the amount of fungicide is relatively low and may only help to reduce seed bed losses of otherwise good quality seed. *Pythium* tip blight occurs as a result of inclement weather and no control is warranted.

Rhizoctonia is encouraged by consolidated soil surface but only occurs at higher than average temperatures. Again, losses are seldom and avoidance of poor soil conditions reduces the risk of infection.

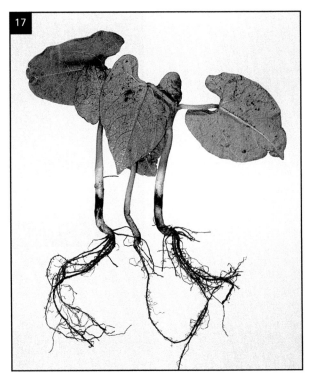

17 Root rot (*Rhizoctonia solani*) of Phaseolus beans.

BLACK ROOT ROT:
Thielaviopsis basicola

18 Field symptoms of black root rot (*Thielaviopsis basicola*) in peas. (Courtesy of PGRO.)

Host Crops
Peas, green beans, dry beans, broad beans, field beans.

Symptoms of Infection
Usually, plants which have begun to flower develop signs of stress, the foliage becomes pale, and growth is stopped (**18**). The root system and stem base is black and this may include all or parts of the fibrous adventitious rooting system (**19, 20**). Plants are affected in patches or larger areas, and very often soil conditions may be good with no obvious signs of compaction or other stress-producing condition.

Economic Importance
The disease is relatively uncommon in Europe and the USA although, in warmer conditions, it can be found a little more frequently. Recently, in the UK, France, and Austria disease levels have been higher than in the past and this has often been associated with the cropping of both peas and *Phaseolus* beans. Yield loss is always associated with poor plant growth and the amount of loss is in proportion to the area infected.

19 Black root rot (*T. basicola*) of Phaseolus beans.

20 Root discolouration of black root rot (*T. basicola*) on a pea plant.

Disease Cycle

Thielaviopsis basicola has a wide host range including large seeded legumes, carrots, ornamentals, and other broad leaved species. The fungus produces chlamydospores which are cigar-shaped structures of cylindrical cells which break up and remain in the soil for many years. Once established in the root system, the fungus spreads throughout the tissue, producing the chlamydospores in the epidermal layer.

Prevention and Control

Close cropping of peas and green beans should be avoided. Soils of high pH are also thought to favour chlamydospore survival. There are no chemical or varietal means of control and so avoidance of fields known to have had a history of the disease is the only way of preventing yield losses.

SECTION 3

Pests of Seedlings and Young Plants

- WIREWORMS

- BEAN SEED FLY (SEED CORN MAGGOT)

- PEA AND BEAN WEEVIL

- FIELD THRIPS

- LEATHERJACKETS

WIREWORMS:
Agriotes spp.

Host Crops
Peas, field beans, broad beans, green beans.

Symptoms of Infestation
Wireworms are the larvae of click beetles. Damage is seen in early spring to early summer. Seedlings are damaged by the wireworms when they bite through the stem or tunnel through seeds (**21**). Depending on the severity of damage, the young plants may not survive. Older plants are usually more able to withstand attack. Signs of injury include narrow holes through the seed or stem at, or just below, soil level. Damaged seedlings may wilt and die along short runs of rows or in patches where the pest has aggregated.

Economic Importance
Usually attacks are sporadic and localized to small areas of the crop. Some plant loss is acceptable but where the infestation is high, then high plant losses occur.

Pest Cycle
In early summer, the beetles lay eggs singly or in small clusters just below the surface of grassy or weedy ground. The adults are elongate beetles, black to brown in colour, and are 6–12 mm long (**22**). They are so-called because of the sound made when jumping. The larvae are very slow growing and can feed on crops or debris for up to 5 years before they mature. In the middle of summer, the larvae, which are thin, yellow and smooth bodied and up to 25 mm in length (**23**), move down deeper in the soil where they pupate and emerge in summer or overwinter as adults.

Prevention and Control
Ploughing of grassland in early spring should be followed by further soil cultivations to expose the larvae to predation or desiccation. There is little opportunity for controlling wireworms once damage has commenced, and re-drilling is not a wise option where the pest remains in the soil.

21 Wireworm (*Agriotes* sp.) larval damage to pea seedlings. (Courtesy of PGRO.)

22 Click beetle (*Agriotes* sp.).

23 Wireworm (*Agriotes* sp.) larva.

BEAN SEED FLY (SEED CORN MAGGOT): *Delia platura*

Host Crops
Green beans, dry beans, peas, broad beans.

Symptoms of Infestation
Seed of late-planted peas or beans is attacked during germination. Larvae from eggs laid in the soil by the adult feed on decaying vegetable matter and also tunnel into freshly imbibed seeds (24). The larvae feed within the cotyledons and damage the developing plumule and root. Tunnelling can take place inside the stem and the growing point can be damaged (25), resulting in a 'baldhead' symptom, where the stem elongates but there is an absence of terminal leaves (26). In peas and Vicia beans, secondary shoots can develop from the seed, but in Phaseolus beans, the seedlings fail to compensate for the damage. Severely damaged seeds fail to produce a seedling and decay before emergence. Damage is often noted in patches in a field as the flies tend to aggregate before egg laying. Late-cultivated fields which contain large amounts of green weed vegetation or fresh crop debris are more likely to be infested.

Economic Importance
Bean seed fly is a very common pest and can be found in most temperate countries affecting a wide range of large seeded crops including peas, beans, marrows, lupins, maize, and soya beans. In severe infestations, large numbers of seedlings can be lost, severely affecting the plant population. This may result in re-drilling and subsequent loss of production of high value vegetable crops early in the season.

Pest Cycle
Adults (27) are attracted to freshly disturbed soil which contains debris from the previous crop, freshly

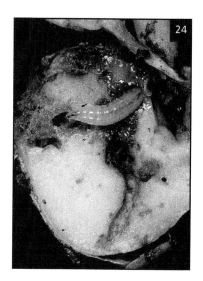

24 Bean seed fly (*Delia platura*) larva in damaged pea seed.

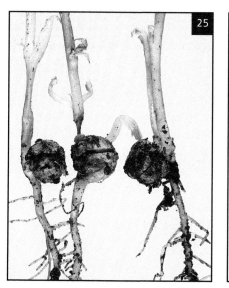

25 Bean seed fly (*D. platura*) larval damage to pea seedlings.

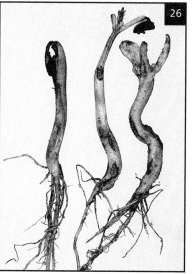

26 Bean seed fly (*D. platura*) larval damage to Phaseolus bean seedlings.

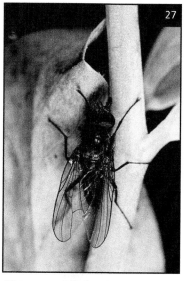

27 Bean seed fly (*D. platura*) adult.

cultivated weeds, or high levels of organic matter such as farmyard manure. Eggs are laid on the soil surface and after a few days larvae hatch and feed on the vegetable material or newly planted seeds. After feeding for 10–14 days, the larvae pupate (**28**) and emerge as a new generation of adults which fly to suitable egg laying sites. There can be several generations each season, the first occurring from late spring and continuing until early autumn.

Prevention and Control

Plantings made from late spring until late summer are more susceptible to damage. Seedbed preparation should ensure that any weed growth has died down before cultivation, or spring cultivations made to reduce the weed population before planting. Runner beans can be planted under polythene or seeds pre-germinated in containers under cover, prior to transplanting for small-scale production. Insecticidal seed treatments are available for peas and beans in some countries and good protection can be obtained where these are used. In some cases, a soil application of insecticide may be used immediately after sowing, but the level of control is very much less than with seed treatment.

PEA AND BEAN WEEVIL:
Sitona lineatus

Host Crops

Peas, field beans, broad beans.

Symptoms of Infestation

Early sown spring peas and broad and field beans are usually the most susceptible to attack by pea and bean weevil. Feeding damage by the adult weevils is characterized by semi-circular notches around the leaf edges of newly emerged seedlings, and the damage is seen as soon as the first leaves begin to expand (**29, 30**). In cool growing conditions, the loss of leaf area can outstrip new growth. The adult can continue feeding for some time on newly developing leaves, but often the young plants will grow away from the attack. Larvae (**31**), produced from eggs laid by the feeding adults, feed below ground on the nitrogen-fixing root nodules (**32**), thereby reducing the available nitrogen and allowing invasion by soil-borne, root infecting fungi. The pest is present in most areas where peas or beans are grown including the USA, Europe, Russia, and Australasia.

Economic Importance

Seedlings and young plants may have a considerable area of leaf tissue eaten but the plants do not often appear to be permanently damaged, except where continual feeding occurs in periods of slow growth. Damage to the root nodules, however, can have an effect on yield, particularly in crops grown as dry harvesting, when the seed weight may be reduced due to nitrogen deficiency. This is particularly noticeable when plant damage occurs in conjunction with other factors such as drought or poor soil conditions, which result in plant growth stress. Experiments have shown that leaf damage alone does not result in yield loss but where larval attack is severe, yield can be reduced by about 25%.

Pest Cycle

The adult weevil is beetle-shaped and about 4–5mm long. Its colour varies between grey-brown to a sandy or a dark brown, with faint striping along the length of the wing cases (**33**). It has a short rostrum

28 Bean seed fly (*D. platura*) pupa.

but conspicuous elbowed antennae. It shelters under soil clods once in the crop but, while feeding, the adults climb freely over the foliage.

The adults overwinter in grassy field edges of the last host crop or in hedge bottoms or ditch sides. As temperatures increase in early spring, weevils migrate to newly emerged pea or bean crops or to established winter sown legumes. The adults move across the soil surface when daytime temperatures exceed 12°C, but will fly when these rise to above 18°C. Often damage is first noted along the headlands before the adults have migrated fully into the crop. Eggs are laid very shortly after arrival into the host crop and are washed down into the soil by rain. The creamy white larvae have dark brown heads and they feed within the root nodules for 6–7 weeks. After maturing, the weevil larvae pupate in the soil (**34**) and emerge as adults in the middle of the summer (**35**). They often feed on any green leaf area present in the crop before moving to the overwintering sites. There is only one generation each year.

29 Leaf notching caused by adult pea and bean weevil (*Sitona lineatus*) on pea.

30 Notching damage to Vicia bean leaf caused by adult pea and bean weevil (*S. lineatus*).

31 Larvae of pea and bean weevil (*S. lineatus*).

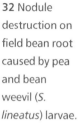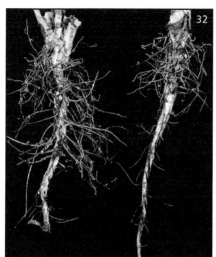

32 Nodule destruction on field bean root caused by pea and bean weevil (*S. lineatus*) larvae.

33 Pea and bean weevil (*S. lineatus*) adult.

Prevention and Control

The weevil is one of the most common pests of peas and beans. Peas growing near to Vicia beans are particularly at risk, but attacks can occur wherever the crops are grown. Newly emerging crops growing slowly in cold conditions should be examined for leaf injury, and an insecticide applied before wide scale damage occurs. These sprays are aimed at reducing leaf damage but also disrupt the egg laying period until such time as the plants have become well established. A second spray applied 10–14 days afterwards, improves the level of response.

A monitoring system based on traps with lures containing the *Sitona lineatus* aggregation pheromone has been developed, to identify the time at which peak migration from the overwintering sites is about to occur. The system is available commercially in the UK. Susceptible crops can then be treated in good time and later sowings of peas or Vicia beans may be timed to avoid this period of peak activity. In Europe, insecticidal seed treatments are available which provide excellent control of both adult feeding and larval damage.

34 Pupating larvae of pea and bean weevil (*S. lineatus*).

35 Pea and bean weevil (*S. lineatus*) adult. (Courtesy of Holt Studios/Duncan Smith.)

FIELD THRIPS:
Thrips angusticeps

Host Crops

Peas, field beans, broad beans.

Symptoms of Infestation

Damage is more severe in dry cold springs. Early spring sown peas or beans, growing on calcareous soils containing a high proportion of stone, are susceptible to thrips damage. The shoots of newly emerging seedlings are pale and distorted and growth appears retarded. As the leaflets expand, they are puckered and leathery with small translucent spots which can be over the leaf surface (36). Damaged leaves of Vicia beans develop a rusty under surface (37). Thrips can be found within the developing leaves of the growing point. Occasionally, in peas, damaged plants fail to recover and produce a rosette of basal developed shoots, the leaves of which continue to show a degree of distortion and the plants remain stunted often in discrete patches over the field. This condition is physiological and has been termed as pea dwarfing syndrome or nanism (38).

Economic Importance

Early sowings of peas or beans are often slow to grow away from the damage and can become further damaged by birds. In severe instances where the plants have developed dwarfing syndrome, flowers are not formed or are poorly developed and the yield of the crop is affected if significant areas of plants suffer in this way.

Pest Cycle

Thrips angusticeps can feed on a range of crops including brassica species. They overwinter in the soil as short winged, flightless adults and feed on seedlings in early spring. The adults are tiny, black, elongated insects about 1–1.5 mm in length (39). They produce orange or yellow coloured nymphs during feeding periods, and these can be found on the plant together with the adults (40). Seedlings are infested before they emerge from the soil. After

36 Feeding damage to pea caused by field thrips (*Thrips angusticeps*).

37 Feeding damage to field bean caused by field thrips (*T. angusticeps*).

38 Patches of nanism following field thrips (*T. angusticeps*) damage to peas. (Courtesy of PGRO.)

39 Field thrips (*T. angusticeps*) first generation wingless adult.

40 Field thrips (*T. angusticeps*) nymph.

feeding on the leaf surface, the thrips produce a winged generation which leave the crop from early summer onwards (41); however, some may remain in the soil until the following spring.

Prevention and Control

Stony soils, where brassica crops are grown as part of the rotation, are more likely to contain high populations of thrips. Because damage is seen at the time of seedling emergence, control is difficult, but often plants outgrow the initial effects as soon as temperatures increase, particularly after rainfall. Where damage is experienced each year, prompt treatment of newly emerging crops with a systemic insecticide prevents further damage; however, most contact insecticides are of little value when the insects are protected by the developing leaves. Insecticidal seed treatments are used in some European countries and these give excellent control of thrips.

41 Field thrips (*T. angusticeps*) winged adult.

LEATHERJACKETS: *Tipula* spp.

Host Crops

Peas, field beans, broad beans, green beans.

Symptoms of Infestation

Leatherjackets are the larvae of the crane fly. They live in the soil, feeding on grass roots and other crop debris. Any crop grown in newly ploughed grassland or in fields or stubbles infested with grass weeds the previous autumn may be attacked. In spring, they feed on the newly germinating seedlings and underground stems. The larvae chew a rough-edged cut through the stem just below soil level, causing the plant to collapse. Damage is often concentrated in small scattered areas of the field associated with the distribution of the pest. Damaged plants of peas and Vicia beans may produce secondary stems but these are often weak and fail to survive. Green beans are not able to compensate for stem damage. The pest is abundant in Europe, northern Asia, Canada, and northern USA. It has a wide range of host crops.

Economic Importance

When infestation is high, plant loss is significant. There is also a high risk to re-sown crops if no control of leatherjackets has been possible.

Pest Cycle

Adult crane flies are slender insects with narrow wings and long fragile legs. In the UK, they are known as 'daddy long-legs' (42). The thorax is segmented with faint stripes and a dusky grey colour. Adults emerge from grassland in early summer and are abundant in late summer and early autumn. Eggs are laid in late summer. They are placed just below the soil surface in small batches. After 2 weeks, the eggs hatch and the larvae feed on grass roots. In spring, the larvae or leatherjackets are fully grown. They have a dull, brown-grey coloured body about 30 mm long. The body is slightly tapered with a tough skin, hence the name (43). They are fully developed by early summer and pupate in the soil, emerging as adults in the same summer.

Prevention and Control

Fields with a history of grassland or long-term pasture are more likely to be infested. Where grassland is cultivated in the autumn, the sods should be examined for leatherjackets and treatment applied to the soil before planting the crop. There are very few insecticides available for leatherjacket control. Avoiding infested field s is the most effective means of prevention.

42 Crane fly (*Tipula* sp.).

43 Leatherjacket (*Tipula* sp.): larva of a crane fly.

SECTION 4

Fungal and Bacterial Diseases

- ALTERNARIA LEAF SPOT
- COMMON ROOT ROT
- BEAN LEAF AND POD SPOT
- PEA LEAF AND POD SPOT
- BOTRYTIS POD ROT
- CHOCOLATE SPOT
- ANTHRACNOSE
- POWDERY MILDEW
- FUSARIUM WILT
- PEA FOOT ROT
- FUSARIUM ROOT ROT (FUSARIUM YELLOWS)

- BEAN FOOT ROT
- BEAN DOWNY MILDEW
- PEA DOWNY MILDEW
- ASCOCHYTA LEAF SPOT
- HALO BLIGHT
- PEA BACTERIAL BLIGHT
- WHITE MOULD, SCLEROTINIA
- STEM ROT
- BEAN RUST
- BEAN RUST
- COMMON BLIGHT

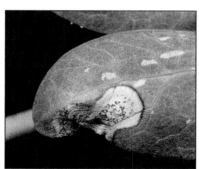

ALTERNARIA LEAF SPOT:
Alternaria alternata

Host Crops
Green beans, dry beans, runner beans.

Symptoms of Infection
The disease is likely to occur on the older trifoliate leaves of beans, especially during prolonged periods of wet cool weather. Small, brown, irregular-shaped spotting occurs on the leaf surface, and the spots develop into larger grey-brown, round lesions containing concentric rings. The spotting tends to occur between the major leaf veins. As they become larger the lesions become more angular, and these can then coalesce causing larger areas of the leaf to die back (**44**). Pod spotting can occur in the form of red-brown flecking over the surface (**45**).

Economic Importance
The disease has caused problems in processing beans in the USA and in Europe, and is particularly destructive if pod lesions develop as this results in blemished produce.

Disease Cycle
The principal pathogen, *Alternaria alternata,* is common in all situations. Spores are produced on the surface of diseased lesions and can be spread easily by rain. However, the leaf surface needs to be wet for at least 24 hours before the spores are able to germinate and infection to develop. The disease is not seed transmitted.

Prevention and Control
Control measures are seldom warranted as the conditions which favour infection are not usual during the bean growing season. High plant populations and excessive rain or irrigation are conducive to disease development.

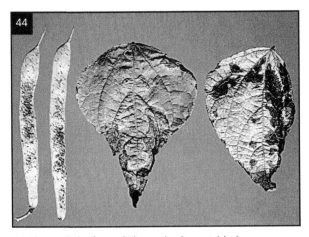

44 Alternaria leaf spot (*Alternaria alternata*) lesions on leaves and pods. (Courtesy of R. Rand.)

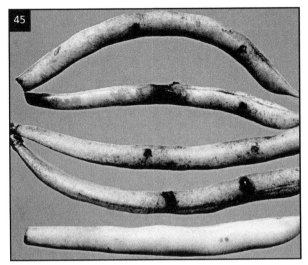

45 Alternaria pod lesions (*A. alternata*). (Courtesy of R. Rand.)

COMMON ROOT ROT: *Aphanomyces euteiches*

Host Crops
Peas, green beans, dry beans, field beans, broad beans.

Symptoms of Infection
Peas can develop symptoms from any time after the first two or three leaf nodes have been produced but, more often, it is at flowering time that the plants show the most obvious effects. Plants are stunted in patches that can be small to extensive over the field (**46**). These areas often coincide with wetter areas of soil or where drainage has been impeded by cultivation compaction. Plants are severely stunted, the root system is decayed, there is an absence of viable nitrogen-fixing nodules, and the plants die from the base upwards. Often, the epidermis of the roots strips away during an examination to leave the vascular strands exposed (**47**).

Economic Importance
This is a very common disease in the USA, UK, particularly Scotland, Scandinavia, and Europe where it has become a major disease in northern France. Crop loss is total in infected areas. The disease is able to survive for many years in the soil, and infects plants that are suffering from root stress brought about by waterlogging or soil compaction, or both.

Disease Cycle
The fungus is soil-borne and survives as thick-walled oospores for many years. Survival is favoured in moisture retentive soils. Infection is facilitated by damaged roots from soil stress, and the oospores produce motile zoospores which invade the damaged root tissue. Mycelial development is confined to the root epidermis which eventually breaks down, resulting in loss of nutrient to the growing plant. Oospores are produced in abundance in the root tissue and remain in the soil after ploughing.

46 Common root root (*Aphanomyces euteiches*) infected pea crop.

47 Common root rot (*A. euteiches*) on pea plants.

Prevention and Control

Some varieties are less susceptible to infection than others, but many new breeding lines are not yet available for commercial production. The disease is favoured by wet conditions and poor drainage. Production of a plough pan should be avoided by cultivating in dry conditions and using minimum tillage in the spring before planting. Where the disease has been a severe problem in peas, then the land should be rested for at least 10 years, to allow the population of oospores to fall to safer levels.

Later planting in susceptible fields and moderate irrigation during the early part of the season will help to reduce the severity of infection. The use of green manure crops such as oats, brassicas and Sudan grass have shown some benefit in the USA. The planting of such crops as alfalfa or clovers will not decrease inoculum levels, but will actually increase or sustain inoculum level. A soil test is available in many countries, which enables soil to be indexed in advance to identify high risk situations.

BEAN LEAF AND POD SPOT:
Ascochyta fabae = Didymella fabae

Host Crops
Field beans, broad beans.

Symptoms of Infection
The first sign of infection appears as small discrete round lesions on the young leaves of newly emerged beans (48). These become visible on winter sown beans in the early spring, or later on in the summer on spring sown crops. The lesions vary in size from small initial infection sites of 1 mm up to larger diameter spots of up to 5–8 mm, and have a light brown to grey coloured centre which is often surrounded by a darker grey to black area, which may be elongated or oval in shape. Within the lighter centre, numerous dark pinpoint-sized protuberances, which are the pycnidia of the asexual stage, *Ascochyta fabae*, appear as the lesions become older. The spotting can extend over the leaf surface and become more numerous. Lesions can develop on the stems, which may be weakened and break off. The stem becomes marked with a red-brown discolouration in the areas of the lesions (49). Dark sunken lesions are produced on the pods following a spread of infection by rain splashed spores (50). Deep seated infection can result in discoloured seed.

Economic Importance
In wet seasons, the disease can spread by rain splashed spores over the surface of the plant and the immediate surrounding plants (51). Plant death may

occur in severe instances but, usually, the stems are weakened and break off or result in the crop lodging as pod development progresses. Pods may be severely infected and shrivel before seed development has been completed, and the seeds may be blemished or be undersized. In dry harvested field beans, blemished or diseased seed may be unsuitable for quality human consumption markets, and in the vegetable broad beans, the pod spotting makes the produce unsaleable for the fresh picked market; blemished beans in vining broad beans can lead to crop rejection by the food processors.

Disease Cycle
A. fabae is primarily seed-borne, and develops on seedlings shortly after emergence. Pycnidia produced within the disease lesions contain spores which are released by the action of falling rain drops or irrigation sprinklers. Spores are splashed onto surrounding tissue and, after a short period of humid weather conditions or prolonged leaf wetness, the infection sites develop into larger lesions on the leaves and stems and pods. Internal systemic infection may also take place, resulting in deep seated internal seed infection. Spores can also be transmitted by rain splash from lesions produced by infected volunteer plants from adjacent fields which carried the previous years crop. A further source of infection from air-borne spores is also likely, where the perfect stage of *Ascochyta fabae*, *Didymella fabae*, develops on the

surface of crop debris over the autumn and winter period. The spores are much smaller than those of *Ascochyta fabae* and are, therefore, believed to be able to be transported greater distances than would otherwise occur due to rain splash. It is this means of transmission that makes the control of *Ascochyta* difficult in Vicia beans.

Prevention and Control

The principal cause of infection is from seed-borne sources, and the use of healthy seed is an important first step in reducing the risk of infection. Control of seed-borne infection by fungicidal seed treatments is only partially effective with existing fungicides. Siting beans well away from the previous crop will also help to isolate them from wind or rain splashed spores from volunteer beans, but it is not known how far wind-blown *Didymella* spores can be transported. Ploughing harvested crops of beans will bury the trash, making the development of *Didymella* less likely.

Seed treatment will have a suppressive effect on low levels of seed-borne infection, but foliar applied fungicides also have only limited ability to control infection once it has become severe. However, some fungicides are able to reduce both the rate of infection and the development of seed-borne infection, if applied to the crop early before the disease becomes established. There are now some resistant winter sown varieties used in Europe.

48 Ascochyta leaf spot (*Ascochyta fabae*) leaf lesions.

49 Ascochyta leaf spot (*A. fabae*) leaf and stem lesions.

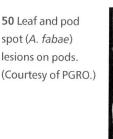

50 Leaf and pod spot (*A. fabae*) lesions on pods. (Courtesy of PGRO.)

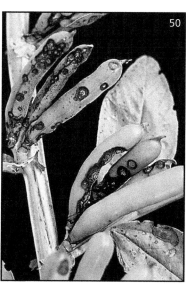

51 Ascochyta leaf spot (*A. fabae*) infection in young Vicia bean plants.

FUNGAL AND BACTERIAL DISEASES

PEA LEAF AND POD SPOT:
Ascochyta pisi, Mycosphaerella pinodes, and *Phoma medicaginis* var. *pinodella*

Host Crops
Peas.

Symptoms of Infection
Three closely related fungi comprise the *Ascochyta* complex, and can cause leaf and pod spot in peas. All are seed-borne and can survive on crop debris, although *Mycosphaerella pinodes* and *Phoma medicaginis* can also be soil-borne. Symptoms of each pathogen are different, but *A. pisi* produces the most distinctive type of leaf and pod spot (52, 53). Infected seeds give rise to infected seedlings and the symptoms appear shortly after emergence, particularly if the weather has been wet. Brown to grey coloured lesions develop on the stem, leaves, or stipules. They are oval or rounded in shape, 2–4 mm in size and, in the centre areas of the lesions, darker coloured, pinprick-sized fruiting bodies are scattered over the surface (54, 55). As the plant develops, the stems or growing points may collapse as the lesions encircle the stem (56), or a more general leaf spotting develops on the foliage. When spores penetrate the pod wall, deeper, brown coloured lesions develop, and these may cause blemishes on the surface of the developing seeds.

Symptoms caused by *Mycosphaerella pinodes* develop later in the season, usually from the time of

52 Ascochyta pod spot (*Ascochyta pisi*) pod lesions.

53 Leaf and pod spot (*Mycosphaerella pinodes*) pod lesions.

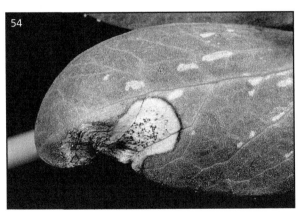

54 Ascochyta leaf spot (*A. pisi*) leaf lesion.

55 Ascochyta leaf spot (*A. pisi*) leaf infection.

flowering, but this may be earlier if the crop is irrigated earlier on. The lesions are smaller and more numerous than those caused by *A. pisi*, purple-brown in colour, and affecting all parts of the plant (57, 58). Often, the stems become black, especially on the lower parts. The disease develops more readily in wet seasons following warmer temperatures and a prolonged period of leaf surface wetness.

Phoma also produces smaller sized leaf lesions, but the more common symptom is a blackened girdling of the stem base resulting in premature senescence, stunting, and death of plants in patches in the crop.

Economic Importance

Infected seeds may fail to emerge, thereby reducing the plant population. *Mycosphaerella* infected seedlings become distinctly blackened at the area of the hypocotyl and decay before emergence. *A. pisi* infection can reduce seedlings after emergence but the general leaf loss and pod spoilage results in yield and quality loss. *Mycosphaerella* infection may cause the plants to lodge before harvest, making harvesting difficult. The spotting of the pods also reduces the quality of the fresh produce (59). The disease results in serious yield loss in wet conditions.

56 Ascochyta leaf spot (*A. pisi*) stem lesion.

57 Leaf and pod spot (*M. pinodes*) lesions.

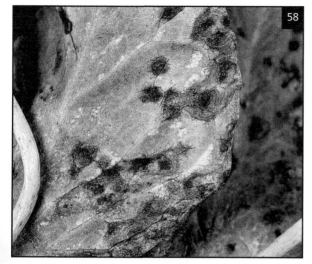

58 Leaf spot (*M. pinodes*) lesions.

59 Leaf and pod spot (*M. pinodes*) lesion on mange-tout peas.

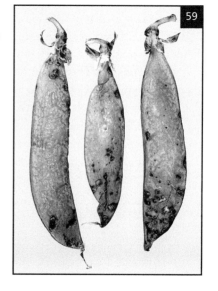

FUNGAL AND BACTERIAL DISEASES

Disease Cycle

All three species of fungi are seed-borne. Infected seedlings produce lesions where pycnidia develop. These pycnidia contain the pycnidiospores which are released during periods of wet weather. The spores are splashed by rain or irrigation water onto surrounding tissue where they germinate and penetrate the plant, producing the characteristic lesions. Pod penetration results in the seeds becoming invaded by the fungi. The fungi can all remain viable on crop residues, although *A. pisi* does not produce resting spores and does not survive for longer than a year. Both *M. pinodes* and *P. medicaginis* produce thick-walled chlamydospores which can remain in the soil for several years. The chlamydospores are then able to germinate and penetrate the stems of newly germinating seedlings. *P. medicaginis* has a wider legume host range that *M. pinodes* or *A. pisi*, which includes lucerne (alfalfa) and *Vicia faba* beans.

Prevention and Control

Seed can be tested for the presence of all three fungi using the agar plate method of culturing the fungi from imbibed pea seeds. The use of healthy seed will prevent problems caused by *A. pisi*, and seedling losses can be reduced if *M. pinodes* or *P. medicaginis* infected seed is not used. However, both these pathogens can survive in the soil and so a rotation that allows a nonlegume crop to be grown for four or five seasons between peas, will help to prevent the build up of high soil-borne inoculum levels.

Seed treatment is available in the form of thiabendazole or fludioxynil fungicides, although there are reports of some populations of both *A. pisi* and *M. pinodes* becoming resistant to thiabendazole. Fungicide sprays can be very effective particularly against *M. pinodes*. The application should be made from flowering onwards with the most effective protection being given by a spray at the first pod set stage and repeated 14 days later. Mixtures containing chlorothalonil with either azoxystrobin, vinclozolin, or cyproconazole are commonly used in Europe.

BOTRYTIS POD ROT:
Botrytis cinerea

Host Crops

Peas, green beans, dry beans, runner beans.

Symptoms of Infection

During wet weather, the flower petals may remain adhering to the ends of pods that have just formed. The pods become infected and the fungus produces a grey furry mould over the remains of the petal. The fungus is then able to penetrate the pod and a watery rot develops from the point of contact of the flower (60–64). The fungus produces fruiting bodies made up of conidia and conidiophores, and the infected tissue develops a grey mould from which the conidiophores are released as a cloud when disturbed. The flower petals may also lodge in the leaf axils and grey mould develops as a girdle around the stem at the leaf node (65–67). The tops of the plants may then senesce and collapse. In green beans, stems may become girdled as infection occurs at soil level at the point of contact with the remains of the cotyledon.

60 Botrytis pod rot (*Botrytis cinerea*) on peas.

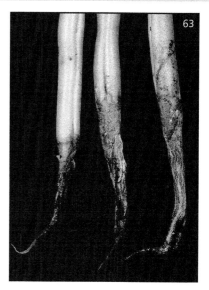

61 Pod necrosis caused by Botrytis pod rot (*B. cinerea*) in peas.

62 Botrytis pod rot (*B. cinerea*) sporulation on pea pod.

63 Botrytis pod rot (*B. cinerea*) on green Phaseolus beans.

64 Sclerotial development of Botrytis pod rot (*B. cinerea*) on green Phaseolus beans.

65 Stem girding caused by grey mould (*B. cinerea*) on peas.

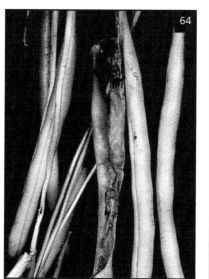

66 Infection site of grey mould (*B. cinerea*) on peas.

67 Infection site of grey mould (*B. cinerea*) in pea leaf axil.

Economic Importance

The loss of pods affects the yield but pod blemishes can make the fresh picked product unsaleable. If infected pods are packed in store prior to shipment, infection can develop on those pea or bean pods which are in contact with the diseased tissue. Peas inside infected pods may also decay and become blemished. Produce for freezing or canning may result in the whole crop being rejected by the processing companies, and in the dry harvested crop, infected seeds become chalky and dry and crumble as they are processed.

Disease Cycle

Botrytis cinerea is a ubiquitous fungus and can be found on virtually all plant species. It is a weak pathogen but a very effective saprophytic mould which can then invade healthy tissue once it is in contact. The disease is favoured by humid conditions and is often associated with vegetable rotting during wet weather. The spores are air-borne and germinate during periods of leaf wetness or high humidity. In peas and green beans, the first site of infection is moribund flower tissue or dying leaves followed by mechanical or pest damage.

Prevention and Control

The disease is favoured by high humidity within the crop. Avoiding a high plant population and encouraging an open crop canopy will assist in allowing the foliage to dry more quickly during the flowering and pod setting period. Overhead irrigation should not be applied during the flowering period. There are no varieties which are resistant, but the afila types tend to be less prone to infection due to the openness of the crop in reducing the risk of a humid microclimate. Hand picked crops should be sorted to avoid pod to pod contact with diseased tissue.

Several fungicides are available which reduce the risk of infection of the pods or stems during wet weather, but the application must be made as a preventative treatment as soon as the first pods can be seen and the majority of flowers have still to fall. Many populations of resistant strains of *Botrytis cinerea* are common and this reduces the range of choice of product.

CHOCOLATE SPOT:
Botrytis fabae, B. cinerea

Host Crops

Field beans, broad beans.

Symptoms of Infection

The earliest stage of chocolate spot infection occurs in the very early spring on leaves of winter sown field and broad beans. Small, round, discrete chocolate coloured spots about 1 mm or more in diameter develop on the lower leaves of the young plants. The lesions may coalesce and form more unevenly shaped grey-brown coloured patches. As the infection develops on the newly developing leaves, the spotting becomes more prolific and lesions are larger in size (68–71). In both winter and spring sown beans, the disease may eventually develop as an aggressive phase which takes the form of larger lesions which may have a zonation pattern in the centres. Streaks develop on the stems which are rusty brown in colour. Pod surfaces are also finely speckled and sometimes bronzed. The plants may defoliate when leaf infection is severe. Chocolate spot is favoured by overcast humid conditions. Production of cuticular wax on the leaf surface is poor under these conditions, and *Botrytis* spores which settle on the leaves quickly germinate and penetrate the leaf developing localized lesions. Occasionally, larger zonate lesions may be found in addition to those caused by chocolate spot. This can be infection caused by *Stemphylium* or *Cercospora* spp. It is uncommon but is favoured by warm damp conditions.

Economic Importance

In summers featuring prolonged wet and overcast periods, the disease can cause premature defoliation and severe yield loss due to a reduction in the photosynthetic area. In beans for fresh market, pod blemishing can result in down-grading of quality and may make the beans unmarketable. Yield losses of around 25% are not uncommon.

68 Early symptoms of chocolate spot (*Botrytis fabae*) on Vicia bean leaf.

69 Established chocolate spot (*B. fabae*) lesions on Vicia bean leaf.

70 Aggressive chocolate spot (*B. fabae*) beginning to defoliate Vicia bean plants.

71 Aggressive stage of chocolate spot (*B. fabae*) on Vicia bean leaf.

Disease Cycle

Although both *B. cinerea* and *B. fabae* can be seed-borne, the main source of infection is from air-borne spores. Both fungi are very common in areas where beans are grown regularly. Field's where beans were the previous year's crop, may contain crop debris and volunteer plants where chocolate spot can overwinter. Spores are produced readily in moist conditions and are transported easily by wind (**72**).

Prevention and Control

Winter sown crops are usually more susceptible to infection as chocolate spot can be present at a low level for most of the winter before developing in wet spring seasons. In spring sown crops, the disease is usually first seen when the crop is beginning to flower. Planting beans well away from the previous years crop or isolating spring sown beans from winter planted crops will help to reduce the infection risk. A deficiency of phosphate in the soil can also result in the crop being more susceptible, and densely planted beans encourage humid microclimate within the canopy, thereby encouraging infection. Some pre-emergence applied residual herbicides, such as simazine, can predispose beans to infection if the plants have taken up some of the herbicide through the root systems.

Fungicides applied at the first sign of spotting will help to prevent rapid infection. In winter sown crops, a spray in the spring at the onset of flowering and repeated 4 weeks later has been shown to be an effective programme. In spring sown beans, a single spray during mid- to late flowering is usually sufficient to control the disease.

ANTHRACNOSE: *Colletotrichum lindemuthianum*

Host Crops

Green beans, dry beans.

Symptoms of Infection

The disease is common in all parts of the world, although damage is more likely to occur in temperate and subtropical areas. Spotting by anthracnose can occur over the leaf surface or on the stems. It is seed-borne, and seedlings produced from infected seed may develop lesions on the stem just as the seedlings are emerging. Leaf lesions are angular in shape, brown-black in colour, with reddish edges (**73**). Small sections of leaf veins may blacken and the central areas of leaves fall out, resulting in a shot-hole effect. Pod lesions are deeply sunken, oval, and dark brown to black in colour. Spores are produced on the surface of the lesions, appearing as small masses of pink to orange specks. Individual plants may be infected but the fungus can be spread by rain splashed spores to surrounding plants (**74, 75**).

Economic Importance

The disease is very destructive and can cause serious losses both in the number of plants killed, thereby reducing yield, and especially by the damage to the pods.

Disease Cycle

The fungus is seed-borne. Lesions developing on the stems of infected seedlings produce spores which are then spread by wind-blown rain or irrigation water to surrounding tissue. Once established, pod lesions develop and the developing seeds become infected by the invading mycelium. Seed harvested from diseased crops is then infected.

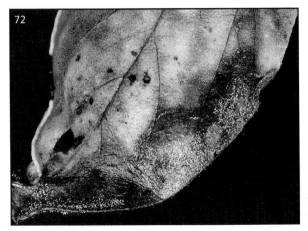

72 Sporulating chocolate spot (*B. fabae*) lesion on Vicia bean leaf.

Prevention and Control

Many commercial varieties of green beans are resistant to the common races of anthracnose, but seed production in dry areas has reduced the level of seed-borne infection significantly. Healthy seed is the main form of prevention; however, systemic fungicides will reduce the risk of infection if present. It is likely that the widespread use of fungicides for *Botrytis* control has kept losses from anthracnose to a minimum in the intensive production areas of Europe and the USA.

74 Anthracnose (*C. lindemuthianum*) on green Phaseolus beans. (Courtesy of P. Miklas.)

73 Vein blackening by anthracnose (*Colletotrichum lindemuthianum*) on green Phaseolus beans. (Courtesy of P. Miklas.)

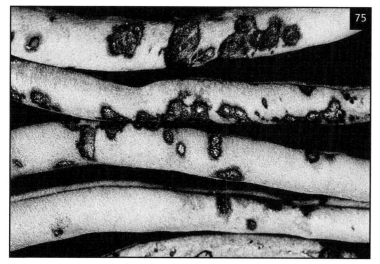

75 Pod lesions of anthracnose (*C. lindemuthianum*) on green Phaseolus beans. (Courtesy of P. Miklas.)

POWDERY MILDEW:
Erysiphe pisi

Host Crops
Peas.

Symptoms of Infection
The disease is favoured by warm temperatures during the day, followed by cool, humid nights. Late sown peas are more susceptible to infection, and it is in these crops that the effects are greatest. Symptoms appear as small irregular areas of powdery white mildew on the upper leaf surface of the lowest leaves and stipules (76). The lesions spread rapidly up the plant and the stem becomes completely covered by the filmy white growth. Pods then become infected and, as the fungus ages, it darkens and produces small black fruiting structures within the lesions (77). The plants may produce a distinctive fishy smell when disturbed and the white spores powder off in clouds during harvest. Severe infection can cause leaves to shrivel and drop off and pods fail to fill.

Economic Importance
Powdery mildew is widely distributed in all areas where peas are grown. Severe infection reduces the pod weight and yield and seeds fail to develop their full potential in infected pods. Occasionally, peas may be hollow if pod infection is severe. The maturity of peas grown for dry harvest may be delayed following infection. Pods grown for the fresh market are blemished and are unsaleable (78). Later sown crops grown to extend the harvesting period, produce very low pod numbers and the harvest period is severely shortened once infection has begun. The unpleasant smell at harvest adversely affects the flavour of processed peas, and the dust at harvest affects the machinery and the operators.

Disease Cycle
The fungus overwinters on crop debris and on alternative host plants such as vetches and other wild legumes. Spores produced on the wild hosts are released during periods of dry windy weather following nights when dew has formed. The spores are transmitted by wind and air currents to peas and infection begins after a very short period under favourable conditions. Seed is not thought to be a source of infection.

Prevention and Control
Varietal resistance to powdery mildew is largely conferred by two single recessive genes, *er* and *er²*. There are several varieties commercially available which are completely resistant to the disease under most field conditions. There is now a complete range of varieties both for the fresh market and for processing which have good agronomic characteristics; hence, the disease should give little problem if the correct choice of variety is made. Fungicides can be used in some situations. Sulphur applied regularly as a foliar spray reduces infection, but protection can be obtained by triazole-based products.

76 Pea powdery mildew (*Erysiphe pisi*).

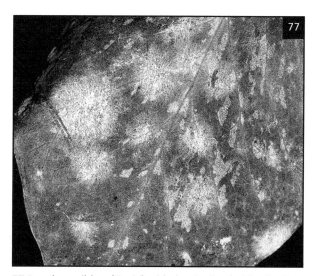

77 Powdery mildew (*E. pisi*) with developing fruiting bodies (cleistothecia).

78 Severe pea pod infection by powdery mildew (*E. pisi*).

FUSARIUM WILT:
Fusarium oxysporum f. sp. *pisi*
Races 1, 2, and 5

Host Crops
Peas.

Symptoms of Infection
Diseases caused by *Fusarium oxysporum* often result in a vascular wilting of the plant during the time of flowering and pod setting. Race 1 of Fusarium wilt is characterized by a stunting of the plant, together with a colour change affecting the whole foliage which, at first, turns a greyish yellow before shrivelling and resulting in the death of the plant. The leaves roll inwards and when the stem tissue is scraped back to reveal the vascular tissue, the colour is yellow to slightly orangey-red which is more pronounced at the nodes. Plants are usually affected in patches in the field (**79, 80**). Race 1 is found in most pea growing areas of the world.

Race 2 of Fusarium wilt is also known as near wilt. Plants tend to be affected either individually or in scattered areas over the field (**81**). Often the root system is severely decayed, but the plant is not so severely stunted as in symptoms caused by root rots. Again, a vascular wilt of the plant occurs but the colour of the foliage is not so grey as in Race 1 and the discolouration of the vascular system is more of a red colour. Often the plants appear to be wilted on one side first before the other, hence the descriptive name of near wilt. Symptoms usually occur during the latter stages of pod set and are often associated with warm weather. Race 2 has been reported in many countries, including the UK and Europe, particularly in Holland, France, and Hungary. The symptoms are often confused with other soil-borne root infecting pathogens.

Race 6 has been reported in the USA and in Scandinavia, although it probably occurs occasionally in Europe and Australasia. The symptoms are very similar to those caused by soil-borne, root infecting pathogens. The roots are discoloured and the plant dies prematurely. Often individual plants become infected in a field, giving a random appearance to the disease.

Other races, including Race 5, have been described in the USA, but symptoms are very similar to the other races, with Race 5 symptoms being similar to those caused by Race 1.

Economic Importance
Infection caused by Race 1 is usually the most damaging to the crop as large areas of the field can be affected. Plants die well before the pods are fully formed and yield loss is total in the affected areas. Because the fungus can survive in the soil almost indefinitely once the disease is established, peas are susceptible to infection for many years afterwards. Races 2, 5, and 6 are not so common but occasionally can cause some yield loss in parts of the fields affected.

Disease Cycle
Fusarium oxysporum f. sp. *pisi* is almost certainly moved from field to field by seed contamination, seed-borne infection, or soil movement on contaminated machinery. Spores and chlamydospores can be transported in irrigation water or on infected plant debris. Plants become infected through root contact with chlamydospores in the soil.

Prevention and Control
Many varieties have been bred with resistance to Race 1, and some are also resistant to the other races. Crop rotation is of little value once the disease is present in the soil, but a rotation which includes peas no more frequently than once in 5 years will help to prevent the disease building up to damaging levels. Where there has been a long history of pea growing on a field, then Race 1 resistant varieties should be chosen as a safeguard.

79 Fusarium wilt (*Fusarium oxysporum*) Race 1 effect on pea crop. (Courtesy of PGRO.)

80 Pea variety susceptible to Race 1 Fusarium wilt (*F. oxysporum*). (Courtesy of PGRO.)

81 Pea crop infected by Race 2 Fusarium wilt (*F. oxysporum*). (Courtesy of PGRO.)

PEA FOOT ROT:
Fusarium solani f. sp. *pisi* and *Phoma medicaginis* var. *pinodella*

Host Crops
Peas, field beans, broad beans.

Symptoms of Infection
The first signs of infection appear from early flowering onwards. Areas, either defined or scattered across the field, begin to develop a pale colouration which later changes to a more distinctive yellowish colour. Plants within these areas are often stunted, the leaves are small and the lower leaves begin to shrivel upwards to the growing point. Pods are very small, few in number, and often the plant dies before the pods reach their potential (82–84). The base of the plant can be brown or black at soil level, where the stem may also develop a strangled appearance. The root system is poorly developed with an absence of nitrogen-fixing root nodules. Roots are often brown and, if the vascular tissue is exposed at the stem base down to the tap root, there is a brick-red colour, especially where *Fusarium solani* is the main pathogen (85).

Economic Importance
Pea foot rot is one of the most common and damaging diseases in peas in most countries of the world. *F. solani* is found universally, although *Phoma medicaginis* seems to be more restricted to the cooler temperate areas where soil moisture is higher during the summer. Plants are often infected in patches which may or may not coincide with

82 Field infection of Fusarium foot rot (*Fusarium solani*) on pea crop.

83 Fusarium foot rot (*F. solani*) on field pea crop.

84 Area of pea crop affected by Phoma foot rot (*Phoma medicaginis*).

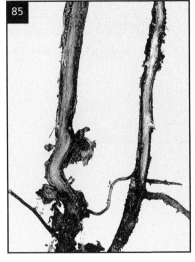

85 Vascular discolouration in pea root caused by Fusarium foot rot (*F. solani*).

compacted or waterlogged soil conditions. Yields are significantly reduced in the severely affected areas and the maturity of the vining crop can be affected as the plants die prematurely, resulting in uneven ripening of the whole crop. Yields of severely affected fields can be reduced by as much as 75% and sometimes crops for harvesting for processing may mature out of sequence, resulting in the crop being left unharvested.

Disease Cycle

The causal organisms of the disease complex are soil-borne fungi, which can survive in the form of thick-walled chlamydospores in field soil for many years. They are often present together in the soil and, in combination, infect the plants resulting in the foot rot symptoms. When a host crop is grown, root exudates stimulate the *Fusarium* chlamydospores to germinate and invade root tissue, especially if the roots are under stress or some physical damage has occurred. Once the epidermis is invaded, *Fusarium* extends to the base of the plant producing fusaric acid, which creates a red precipitate in the vascular tissue. The root epidermis breaks down and any root nodules decay. Chlamydospores produced in the diseased tissue are returned to the soil after the crop has been destroyed. There is also a risk of contamination occurring on harvested seed and this may be an important route by which new fields become infected in time.

In the case of *Phoma*, the stem base epidermal cells are infected and the fungus surrounds the tissue, which breaks down, shrivels, and dies (**86**). *Phoma* is also seed-borne and, during periods of wet weather, pycnidiospores produced at the stem base are splashed up onto the leaves where small dark brown lesions develop. These in turn produce more pycnidiospores which are then transferred to the pods where the lesions penetrate the pod wall and invade the seed. Large populations of both pathogens build up in soil which is frequently cropped with peas. There are several closely related *forma speciales* of *F. solani*, but those which infect peas can also infect *Phaseolus vulgaris* beans and *Vicia faba* beans. In the laboratory, isolates of *F. solani* from peas have also been shown to infect soya beans. *Phoma* is a pathogen of peas, *Vicia* spp. and *Medicago* spp. (lucerne).

Prevention and Control

Newly introduced pea varieties have been developed to be more tolerant to *F. solani*, although no variety has been shown to be resistant in fields where the disease pressure is high. Almost all varieties are highly susceptible to *Phoma* infection. Soil conditions which allow physical damage or growth restriction to the root system should be avoided by reducing the number of cultivations made in the spring before sowing. Plough pans or compaction by wheelings and turnings on headlands also encourage the disease, as does planting peas in wet soils.

Seed should be tested and found to be free of seed-borne *Phoma medicaginis* var. *pinodella* before planting. Seed treatments containing fludioxynil, thiabendazole, or related compounds have been shown to reduce the severity of *Phoma* infection at the beginning of the season, and to be particularly useful in improving the survival of autumn sown peas.

The fungus builds up in the soil over a number of years, but by maintaining a nonlegume cropping interval of 5 or more years, the rate of build-up is minimized and the risk of infection reduced. Peas, green beans, and Vicia beans are all host to the pathogen and should be treated as one and the same crop when planning crop rotations.

A soil test has been developed in the UK and is also used in other European countries to identify those fields where the disease risk is high. The test should be done on fields a year from planting so that alternative crops may be chosen if the disease risk to peas is too high.

86 Pea stem base blackening caused by Phoma foot rot (*P. medicaginis*).

FUSARIUM ROOT ROT (FUSARIUM YELLOWS): *Fusarium solani* f. sp. *phaseoli*

Host Crops
Green beans, dry beans, runner beans.

Symptoms of Infection
Small groups of plants begin to develop pale leaves and are stunted in growth. Eventually the leaves senesce and the plants die prematurely (87). The stem base is discoloured, with black and red streaking extending from the roots to just above soil level. The vascular tissue of the stem base is red-brown in colour, and becomes dry and pithy in time (88).

Economic Importance
Fusarium root rot is very common in all bean growing areas. Large-scale loss of plants results in low overall yields but, where infection is not severe, the plants may be unthrifty and fail to reach their full yield potential.

Disease Cycle
Fusarium is a soil-borne fungus and can survive in fields for many years. Inoculum builds up following successive bean cropping and some isolates of *F. solani* f. sp. *phaseoli* can cross-infect peas and Vicia beans.

Prevention and Control
There is some varietal tolerance to root rot, but most of the commercial varieties for processing are susceptible to some extent. A rotation which includes beans or peas no more frequently than once in 5 years will help to reduce the rate of build-up of potentially damaging populations of the pathogen. Soil consolidation can also predispose plants to infection, as does any situation where root growth is restricted or damaged. Avoidance of compaction under the surface and the lack of plough pan consolidation will reduce the effects of *Fusarium*. There is no means of chemical control and once the disease is established in a field, a break of at least 10 years is necessary to allow the inoculum to fall to a safer level.

87 Fusarium yellows (*Fusarium solani*) on green Phaseolus bean crop. (Courtesy of PGRO.)

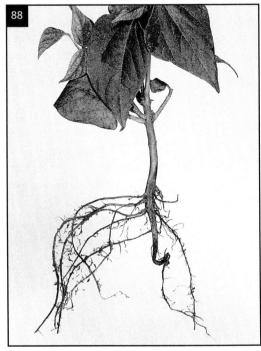

88 Root decay of green Phaseolus beans caused by *F. solani*.

BEAN FOOT ROT:
Fusarium solani, Phoma medicaginis var. *pinodella,* and *Fusarium culmorum*

Host Crops
Field beans, broad beans, peas.

Symptoms of Infection
Beans with blackened stem bases can be found at any time during the late spring or early summer, and are particularly noticeable at the beginning of flowering (**89**). Affected plants are stunted and pale in colour, and the root systems are poorly developed with few, if any, viable nitrogen-fixing nodules. Affected plants occur in patches and are often associated with compacted or waterlogged soil. The internal tissues of the vascular system may be brown or red-brown in colour where *Fusarium solani* is the primary pathogen. *Phoma medicaginis* infection causes a breakdown of the epidermis of the stem base at soil level. The stems may shrivel and the plants develop a strangled appearance. *Fusarium culmorum* infection is characterized by a basal stem rot, which forms a deep lesion just above soil level (**90**). The lesion is black and often contains pink spore masses of the fungus (**91, 92**).

89 Field beans wilting because of Fusarium stem rot (*Fusarium culmorum*) infection.

90 Stem base infection of Fusarium foot rot (*F. culmorum*) on field bean plants.

91 Stem base and root infection of Fusarium foot rot (*F. culmorum*) on field bean plants.

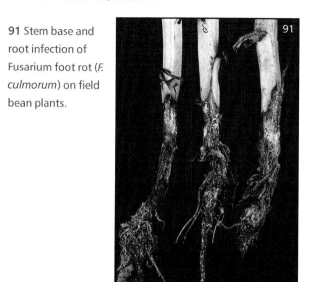

92 Fusarium stem rot (*F. culmorum*) effect on field bean crop.

93 Reseliella gall midge (*Reseliella* sp.) larvae below Vicia stem epidermis.

94 Reseliella gall midge (*Reseliella* sp.) larvae exposed in Vicia stem tissue.

In some seasons, the blackened lesions contain red-pink segmented midge larvae (**93, 94**). These are the larvae of a *Reseliella* midge which lays its eggs in small cracks in the epidermis of the stems of beans, especially where the stems have been damaged by mechanical means or by disease. The presence of midge further encourages *Fusarium* infection and the effects of the foot rot or stem rot become more severe. The effect on plant growth is similar to foot rot, and often the two pathogens are found in infected tissue.

Economic Importance

Loss of plants over large areas of fields significantly affects the yield. Bean seed size is often reduced on surviving infected plants, and uneven maturity of the crop can cause problems with harvest schedules in large-scale broad bean production.

Disease Cycle

All three fungal pathogens are soil-borne and can remain as viable chlamydospores for many years in the absence of a host crop. Infection begins shortly after seedling emergence and is enhanced by the presence of root exudates produced as a result of superficial root damage caused by restrictive soil conditions, compaction, or short-term waterlogging. In some soils, damage by grazing nematodes such as *Pratylenchus* spp. can also facilitate the infection. Damage to the stem above soil level can allow invasion of *Reseliella* midge larvae and these add to the effects of the fungal infection, particularly by *F. culmorum*. The infected crop debris is returned to the soil where the chlamydospores remain until stimulated to infect a newly planted host.

Prevention and Control

Once established in the crop, there is no means of control. Prevention of infection is by good drainage and reduced cultivations made before planting, to avoid compaction caused by repeated passes when creating a seed bed. A high plant population also allows high humidity to develop and this in turn is favoured by the *Reseliella* midge and by infection of *Fusarium culmorum*. Both *Vicia faba* and *Pisum sativum* are hosts to the pathogens, and they should not be grown in the same rotations more frequently than once in 5 years. *F. culmorum* is responsible for ear disease in wheat and crown rot in sugar beet. Beans should not be planted immediately following crops where *F. culmorum* caused noticeable infection the previous year.

BEAN DOWNY MILDEW:
Peronspora viciae

Host Crops
Field beans, broad beans (see also pea downy mildew).

Symptoms of Infection
Although this disease can be found on both autumn and spring planted beans, the most severe infections usually develop on spring sown field or broad beans. Downy mildew of peas is caused by the same pathogen, although it is thought that strains that infect peas are not pathogenic to *Vicia faba* beans. Infection becomes obvious in the late spring when conditions are humid. Young plants can develop symptoms before bud initiation but often the leaves develop the characteristic pale patches on the surface, with grey-mauve fluffy mycelium on the undersides, as they begin to flower. Red-brown flecks or spots can develop over the surface of the lesions and occasionally the edges of the lesions also become discoloured. Scattered infected plants can develop from a soil-borne infection. These are pale coloured and are stunted (**95**). Mycelium developing on the foliage produces an abundance of spores which are then wind-borne to begin secondary infections on surrounding plants. As the disease progresses, the growing points become very pale and distorted and infected tissue can shrivel and die back, reducing the potential for further flower development (**96**).

Economic Importance
The effect of severe infection can be dramatic on the yield of spring sown beans, but on autumn planted crops, the leaf lesions which develop as a result of secondary spread of the spores, are seldom large enough to have any effect on the crop. Downy mildew is one of the most common and damaging diseases of spring beans in Europe, Scandinavia, and other temperate growing areas.

Disease Cycle
Soil-borne oospores of *P. viciae* can remain viable for many years. Seedlings planted in infested soil can develop infection soon after seed germination, but the main source of infection later in the season is from air-borne spores produced from primary infected seedlings. Spores are able to be blown large distances by wind and, hence, the disease is widespread throughout bean production areas. Soil-borne infection is facilitated by cool, moist soil conditions which slow the rate of germination of the seeds. Secondary spread of air-borne spores occurs under periods of high humidity and cool temperatures, although not during rainy periods. Humid conditions are then necessary for spore

95 Seedling downy mildew (*Peronospora viciae*) infection in field bean. (Courtesy of PGRO.)

96 Leaf downy mildew (*P. viciae*) infection in field bean.

97 Sporulation of downy mildew (*P. viciae*) on field bean leaf.

98 Sporulation of downy mildew (*P. viciae*) on field bean leaf.

99 Lesions of downy mildew (*P. viciae*) on field bean leaf.

100 Systemic downy mildew (*P. viciae*) infection of field bean growing points.

germination to occur on the leaf surface (**97, 98**) and infection is more likely to develop rapidly during a period where the temperatures average 10°C and rainfall has occurred 12 hours beforehand. Oospores are produced within the infected tissue and are returned to the soil in the crop debris after harvest. If conditions become warm and dry following an initial infection, disease development is stopped and the plants are able to produce new healthy growth.

Prevention and Control
Beans are known to exhibit differences in varietal susceptibility, and the choice of more resistant types is the most effective means of control. Frequent cropping with beans also allows high populations of soil-borne inoculum to develop.

Chemical control with phenylanylines or fosetyl aluminium is commonly used in the UK, and other fungicides based on dithiocarbamates have also been successful in preventing spread. Treatment is best applied as the disease is first seen and particularly if the crop has just commenced flowering (**99, 100**). Seed treatment with cymoxanil plus metalaxyl R is used occasionally in the UK. Such treatment reduces initial seedling infection and is useful where soils are carrying a high inoculum level.

PEA DOWNY MILDEW:
Peronspora viciae

Host Crops
Peas (see also bean downy mildew).

Symptoms of Infection
Infection occurs from early emergence of the pea seedlings onwards. Individual or small scattered groups of young plants appear stunted and pale in colour. On the underside of the leaves, a grey-mauve coloured velvety growth of mycelium develops and causes pale blotchy patches on the upper surface of the infected leaves (101, 102). Often these primary infected plants die. The disease develops rapidly during periods of cool humid weather. As the plants get older, systemic infection from air-borne spores develops at the growing point, which foreshortens the internodes and produces a stunting effect. Pods develop yellow blotchy patches on the surface, and a profuse cottony growth develops from the inner pod wall.

101 Pea downy mildew (*Peronospora viciae*).

102 Velvety mycelial growth of downy mildew (*P. viciae*) on the underside of pea foliage.

103 Pea pod infection with downy mildew (*P. viciae*).

104 Pea pod blemishes following downy mildew (*P. viciae*) infection.

105 Internal pod wall reaction to downy mildew (*P. viciae*) infection.

Seeds fail to develop and pods fill unevenly (**103–105**). The pod may then become colonized by saprophytic moulds, including *Botrytis cinerea*.

Economic Importance
Downy mildew is the most common disease of peas in temperate areas. The disease is widespread in northern Europe and Scandinavia, New Zealand, and the mid-western and north western states of the USA. It causes significant plant loss early in the season when weather conditions are more favourable for infection. Later infection debilitates the plant and reduces pod and seed set. Pods are blemished and become unsaleable for the fresh market, and peas for processing are often blemished and undersized.

Disease Cycle
Soil-borne oospores germinate following stimulation from germinating pea seeds. The infection progresses to the emerging seedling, and mildew develops on the underside of the leaves. At periods of high humidity, sporangia are produced on the mycelium and released, particularly in the early morning. When these come into contact with new leaf tissue, small lesions develop and later the characteristic velvety mycelium is produced over the surface. Thick-walled oospores are produced prolifically in the pod walls and, when the crop is harvested, these are returned to the soil in crop debris, where they remain viable for many years in the absence of peas.

Prevention and Control
There are several races of pea downy mildew, and most fields with a history of the disease contain mixtures of these. Some varietal tolerance exists and several cultivars of round seeded dry harvest peas are virtually resistant. However, many varieties of wrinkle seeded garden or vining peas are very susceptible and often suffer severe losses when sown early in the season. Cropping rotation is of limited value in preventing disease development, although growing peas more frequently than once in 5 years will significantly increase the chances of infection. Early planting should be avoided where fields are known to have a history of downy mildew.

Seed treatments containing fosetyl aluminium, metalaxyl, or cymoxanil are very effective in protecting seedlings from soil-borne infection, but offer little protection against secondary infection from incoming air-borne spores. Resistance to metalaxyl has developed in several countries including the UK, USA, and New Zealand and seed treatments containing a mixture of active ingredients are now used.

ASCOCHYTA LEAF SPOT: *Phoma exigua* var. *exigua* = *Ascochyta phaseolorum, Ascochyta bolthauseri*

Host Crops
Green beans, dry beans.

Symptoms of Infection
Leaf spotting occurs usually later in the season following a period of wet weather. Symptoms of infection by both fungi are similar. The spots are more or less oval, and brown in the centre which may develop concentric ringing. Small black fruiting bodies are often found in the middle areas of the lesions (**106–108**). Infection may also develop on the pods, particularly those of the runner bean (*Phaseolus coccineus*).

Economic Importance
The disease is seldom serious as it requires cool wet conditions in which to develop. Runner beans are more frequently found to be infected later in the season, when pod blemishes make the beans unsaleable.

106 Leaf spotting caused by *Phoma exigua* on green Phaseolus bean.

107 Vein discolouration in green Phaseolus bean caused by *P. exigua*.

108 Leaf lesions in green Phaseolus bean caused by *P. exigua*.

FUNGAL AND BACTERIAL DISEASES

109 Infection focus of *P. exigua* in green Phaseolus bean crop.

Disease Cycle

Ascochyta is primarily seed-borne, but can survive in crop debris. Runner beans growing in the same field year after year are more susceptible to infection. Spores produced by pycnidia in the lesions are transmitted to surrounding foliage and pods, by rain or water splash (**109**).

Prevention and Control

Long periods of irrigation should be avoided to allow the leaves to become dry during the middle of the day. It may be necessary to rotate runner beans more frequently if the disease becomes more common year after year. Fungicides can be effective in reducing the infection spread during wet weather periods.

HALO BLIGHT:
Pseudomonas syringae pv. *phaseolicola*

Host Crops
Green beans, dry beans, runner beans.

Symptoms of Infection
Early symptoms may be seen on the first or second set of trifoliate leaves. Small brown spots, which may be slightly angular in shape, are surrounded by a pale yellow halo of tissue (**110, 111**). The leaves may become slightly distorted as the centre of the lesion increases in size and the lesions coalesce. The central spot appears water-soaked, with a greasy edge. The infection spreads rapidly during wet weather, and plants collapse as the stems become infected. Pods develop sunken water-soaked lesions, which can have a reddish coloured edge (**112**). As the disease is primarily seed-borne, foci of infection can be found in a crop, and the secondary development of disease affecting surrounding plants can sometimes be seen to follow the direction of the prevailing wind (**113**).

Economic Importance
Halo blight is common in cool wet conditions and affects both green beans and runner beans. Where disease development is rapid, loss of plants may be significant but the effect on the pods is more serious as the spotting blemishes the surface and makes them unsaleable.

Disease Cycle
Halo blight is seed-borne. There is no carry-over of infection in the soil, although infected crop debris may result in infection of later beans planted as a follow-on or second crop in the same season. Bacterial cells are produced on the surface of disease lesions and as 'bacterial ooze' from the pod lesions. They are spread by rain splash or physical movement of contaminated machinery or wet clothing.

Prevention and Control

The main means of prevention is by the use of healthy seed which can be tested before use. However, several varieties of green beans are resistant to halo blight. Copper-based fungicides have been shown to reduce the rate of spread of infection, but these are not always acceptable to processors. The addition of streptomycin sulphate to seed treatments has also been used in the past, but with only moderate success. Removal of infected plants is possible in small-scale production.

110 Foliar symptoms of halo blight (*Pseudomonas phaseolicola*) of Phaseolus bean. (Courtesy of P. Miklas.)

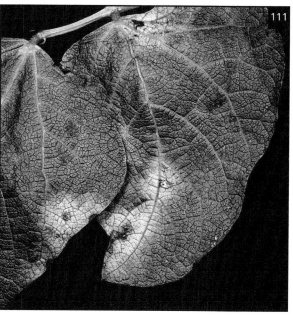

111 Leaf symptoms of halo blight of Phaseolus bean (*P. phaseolicola*).

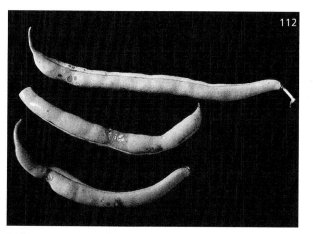

112 Grease spotting lesions on Phaseolus bean pods caused by halo blight (*P. phaseolicola*). (Courtesy of PGRO.)

113 Infection focus of halo blight (*P. phaseolicola*) in Phaseolus bean crop. (Courtesy of PGRO.)

PEA BACTERIAL BLIGHT:
Pseudomonas syringae pv. *pisi*

Host Crops
Peas.

Symptoms of Infection
Symptoms can appear at any time during the growing season, especially where the plants have been physically damaged by pest, machinery or, more commonly, frost or hail. Small, light brown, irregular-shaped lesions can develop on the leaves or stem (**114**). These may be confused with *Ascochyta* spp., but the characteristic feature is the water-soaked or greasy edge to the lesions. These lesions may elongate along the stem and, if the plant has been damaged by frost, the lesions extend in a fan shape, with necrotic streaks extending outwards along the stipules and leaves from the stem node (**115**). Pods may develop greasy spots on the surface which develop into sunken craters, eventually contaminating the developing seeds (**116**). Where plants have been severely damaged by frost and the weather conditions become very wet, the disease develops rapidly and plants may collapse and rot away. In drier conditions, the disease can become quiescent and infected leaves shrivel with no further obvious loss to the crop. Autumn sown peas are more susceptible to infection, as late spring frosts can occur as the plants are lush and at the point of flowering.

Economic Importance
The disease can be serious in autumn sown crops, or crops grown under overhead irrigation. In Europe and New Zealand, where autumn planting in some areas is common, the disease can become very severe and crop loss is not uncommon. In dry conditions, the disease is not severe enough to cause loss. In spring sown peas, significant damage is uncommon, unless there has been damage by frost or hail.

Disease Cycle
Pea bacterial blight is seed-borne but may survive on infected crop debris over one winter. Infected seeds can give rise to seedlings which show no symptoms until physically damaged. Systemic infection can take place, whereby the seeds become infected internally with no visible symptoms, but there is only a very low rate of multiplication in these conditions. Where crops develop symptoms during pod set, then the produce can become highly infected. There are seven races of pea bacterial blight. Race 2 appears to be common in spring sown varieties, while Races 4 and 6 are found in winter sown peas. In the USA, Race 4 seems to be the common race in spring peas.

Prevention and Control
No varieties are resistant to all seven races, although several are resistant to the common Races 2 and 4. There is no chemical means of control and so seed health and good crop hygiene are the only means of control. Seed can be tested for the presence of blight, and only healthy seed should be used where there is high risk of late frost or winter damage. If infected crops are harvested, the seed should be isolated from any healthy seed stocks, as contamination can occur from infected crop debris or dust either in the harvesting and handling equipment or the store itself. Planting crops in areas where frost is likely to occur in the spring should be avoided and a crop rotation maintained which allows a reasonable break from peas.

114 Pea bacterial blight (*Pseudomonas syringae*) in pea crop.

115 Fan-shaped lesions of bacterial blight (*P. syringae*) on pea.

116 Pod infection of bacterial blight (*P. syringae*) on pea.

WHITE MOULD, SCLEROTINIA:
Sclerotinia sclerotiorum

Host Crops
Peas, broad beans, green beans, dry beans, runner beans.

Symptoms of Infection
Individual plants or small groups may be infected in discrete areas over the field. Infection is usually noticed in early to mid-summer, particularly when the weather conditions have been warm and wet. Infected stems become covered with white mycelium and the stems may collapse in a watery soft rot (**117–122**). The stems are often bleached as they desiccate and the upper plant parts wilt and die. Infected stems and pods may also contain black, elongated resting bodies (sclerotia), which may develop on or within the diseased tissue (**123**).

Economic Importance
Infection by *S. sclerotiorum* is common in some areas, particularly where the previous cropping contained other host crops. *S. sclerotiorum* has a very wide host range, which includes vegetables, potatoes, linseed, oilseed rape, sunflowers, and soya beans. The disease spreads rapidly in warm wet conditions. As well as plant loss, the produce may be soiled following infection and the sclerotia can contaminate the harvested produce of both beans and peas, where they are difficult to remove in the processing factory.

Disease Cycle
Sclerotia can remain in the soil for several years. When brought to the surface during cultivations, the sclerotia produce small cup-shaped apothecia, which release spores into the air. Spores may adhere to stems where they have been damaged or they may colonize moribund flower petals before invading the stem tissue. Infection is favoured by wet and warm conditions, and by densely planted crops.

Prevention and Control
A rotation of 4 or 5 years in the absence of alternative host crops will help to prevent the disease occurring in most situations. Where the disease is expected, a preventative fungicide such as vinclozolin or azoxystrobin should be applied during flowering.

117 Stem rot (*Sclerotinia sclerotiorum*) on peas.

118 Sclerotia of stem rot (*S. sclerotiorum*) on peas.

119 Pod rotting caused by stem rot (*S. sclerotiorum*) on peas.

120 Severe stem rot (*S. sclerotiorum*) infection in a green Phaseolus bean crop. (Courtesy of PGRO.)

121 White mould (*S. sclerotiorum*) of Phaseolus green beans. (Courtesy of PGRO.)

122 Pod and stem infection of white mould (*S. sclerotiorum*) on green Phaseolus bean pods.

123 Sclerotia of white mould (*S. sclerotiorum*) developing on green Phaseolus bean stems.

124 Field bean plant infected by white mould (*Sclerotinia trifoliorum*). (Courtesy of PGRO.)

STEM ROT:
Sclerotinia trifoliorum

Host Crops
Field beans, broad beans.

Symptoms of Infection
Winter bean infection is noticed in early spring following invasion of the stems by the fungus. Stems develop a watery soft rot, and then turn black as they decay further before collapsing (**124**). Individuals or small groups of plants may be infected in discrete areas over the field. Infected stems become covered with white mycelium, and black round or oval resting bodies (sclerotia) of the fungus develop in or on the surface of the stem (**125, 126**).

Economic Importance
Infection by *S. trifoliorum* is not common and when disease does occur, the areas of infected plants are usually small and loss of plants is rarely significant.

Disease Cycle
Sclerotia of the fungus can remain in the soil for several years. When brought to the surface during cultivations, the sclerotia produce small cup-shaped apothecia which release spores into the air. Spores may adhere to stems where they have been damaged, or they may colonize moribund flower petals before invading the stem tissue. *S. trifoliorum* requires only moderate temperatures for apothecia development, hence is more likely to infect autumn sown beans in early spring. It also has a very restricted host range of clovers and Vicia beans.

Prevention and Control
A rotation of 4 or 5 years in the absence of alternative host crops will help to prevent the disease occurring in most situations. If infection is noted in early spring, then a fungicide treatment may be used if necessary.

125 Sclerotia of field bean stem rot (*S. trifoliorum*) developing on stem surface.

126 Sclerotia of field bean stem rot (*S. trifoliorum*) developing inside stem.

BEAN RUST:
Uromyces appendiculatus

Host Crops
Green beans, dry beans, runner beans.

Symptoms of Infection
The first sign of infection is the appearance of small red-brown raised spots about 0.5 mm in diameter on the leaf surface. The spots may be surrounded by a narrow, pale coloured zone (**127**). The spots become more numerous after a short time as the infection spreads, and the spotting darkens in colour. Each pustule releases a brown spore mass that can be blown by wind to surrounding foliage (**128**). Pods may also become infected. Infected leaves desiccate and the growth of the plant is stopped. In runner beans, later developing pods are affected and production of new flowers ceases.

Economic Importance
Rust can be very damaging in warm seasons where night time temperatures drop and humidity is high. Yield can be reduced and pod blemishing leads to crop reduction. In runner beans, the harvesting season is truncated and there is a loss of yield and production.

Disease Cycle
Rust survives on crop debris for several months and spores produced in the early summer can be spread onto adjacent crops (**129**, **130**). Once spores are in contact with the leaf surface, a period of high temperature and humidity results in spore germination, and subsequent infection by spores erupting from the pustules is favoured by these conditions.

Prevention and Control
There are many races of *Uromyces appendiculatus* worldwide, and there is little varietal resistance to all races. Crop rotation will reduce the risk of infection from overwintering crop debris, and the destruction of infected plants after harvest will also be of use. Irrigating in the morning to allow the plants to dry before evening will also discourage spore germination. Fungicides such as tebuconazole or azoxystrobin applied as soon as the first pustules can be found are very effective in controlling rust.

127 Phaseolus bean rust (*Uromyces appendiculatus*) symptoms on bean leaf adaxial surface.

128 Sporulating pustules of Phaseolus bean rust (*U. appendiculatus*) on bean leaf abaxial surface.

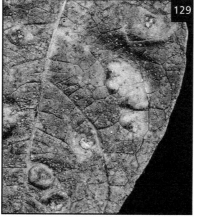

129 Aecia of Phaseolus bean rust (*U. appendiculatus*) on bean leaf adaxial surface.

130 Aecia of Phaseolus bean rust (*U. appendiculatus*) on bean leaf abaxial surface.

BEAN RUST:
Uromyces fabae

Host Crops
Field beans, broad beans.

Symptoms of Infection
Bean leaves become infected from early flowering onwards, but the most serious infections may begin later in the season, following a period of warm days and cooler nights with high humidity. Because of the influence of weather conditions, later maturing, spring planted beans are often more prone to infection, although in some seasons, autumn sown crops can become affected. The first pustules of rust develop on leaves at the middle to lower parts of the stem. The fungus develops as orange to red-brown spots about 1–2 mm in diameter; the leaf tissue immediately surrounding the pustule is pale to yellow in colour, and resembles a halo around the site of infection (**131, 132**). Later the spotting becomes more prolific, the pustules deepen in colour, and masses of brown spores erupt from the now raised pustules (**133–135**). The pustules can develop as 'fairy rings' on the leaf surface. Stems also become infected. Severely infected leaves desiccate and abscise, leaving a partially defoliated plant with poor pod set and development (**136**).

Economic Importance
Rust is common in temperate growing areas, and in areas of large diurnal temperature fluctuation. If the disease develops during the pod development stages or earlier, then the effect on yield is dramatic, with large losses of production in seasons of heavy infection. In some seasons, the disease has more of an effect on yield reduction than the other main fungal diseases, chocolate spot, *Ascochyta fabae*, and downy mildew.

Disease Cycle
Rust can survive over the winter on crop debris or on volunteer crops. Rust pustules can often be found on the lower leaves of volunteer plants surviving in the previous year's fields. However, spore production is encouraged by high humidity and warm temperatures. Spores are released from the maturing pustules and can be blown large distances by wind until they are deposited on a susceptible host crop. Spore germination occurs quickly in the presence of a light film of moisture on the leaf surface. Infection of the rest of the foliage and the surrounding plants follows further production of spores from the new pustules.

Prevention and Control
There are some differences in varietal susceptibility of beans, particularly the European bred, spring sown varieties of field beans. Most varieties of broad beans, however, seem to be susceptible. Cropping in close proximity of the previous year's beans, especially if rust was severe in that crop, should be avoided, and the survival of volunteer bean plants should be discouraged.

Rust can be controlled effectively by a range of triazole fungicides, or by the newer strobilurin fungicides. Treatment made towards the end of flowering, before rust becomes established on the upper parts of the foliage, will provide protection of the crop for a significant period of the season. Often the use of such treatments can be combined to provide protection against chocolate spot.

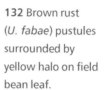

132 Brown rust (*U. fabae*) pustules surrounded by yellow halo on field bean leaf.

131 Early development of brown rust (*Uromyces fabae*) pustules on field bean leaf.

133 Advanced infection of brown rust (*U. fabae*) on field bean plant.

134 Sporulating brown rust (*U. fabae*) pustules on field bean leaf.

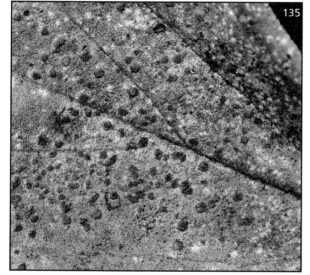

135 Active sporulation of brown rust (*U. fabae*) pustules on field bean leaf.

136 Premature leaf fall in field bean crop caused by brown rust (*U. fabae*).

COMMON BLIGHT:
Xanthomonas campestris pv. *phaseoli*

Host Crops
Green beans, dry beans.

Symptoms of Infection
The disease is favoured by long periods of warm and humid weather. It is less common than halo blight in northern Europe. Symptoms usually occur just prior to flower development, when small water-soaked spots occur on the leaves. These enlarge and become necrotic, with an area of bright yellow tissue surrounding the spots (**137**). The enlarged necrotic areas appear scorched and may later dry up and disintegrate. Pods may also develop sunken water-soaked spots which may be more or less circular, but are often surrounded by a brick red colouration of the edges of the spots (**138, 139**).

Economic Importance
The disease can be very serious in areas where beans are grown under hot and humid conditions, causing loss of plants in infected areas and pod blemishes. Seed from infected pods may also be blemished.

Disease Cycle
Common bacterial blight is seed-borne and the initial focus of infection develops randomly throughout the crop, depending on the initial level of infection in the seed. Bacteria produced from the leaf and pod lesions can be spread by rain splash, wind-blown rain, or irrigation water as well as by mechanical spread by machinery or people.

Prevention and Control
The use of healthy seed is the most effective means of prevention. Once the disease is established in a crop, there is little in the way of effective control except by reducing the time of irrigation and limiting the traffic of people or machinery. A minimum rotation of 2 years will prevent carry-over of infection from one crop to the next.

137 Common blight (*Xanthomonas campestris*) of green Phaseolus beans.

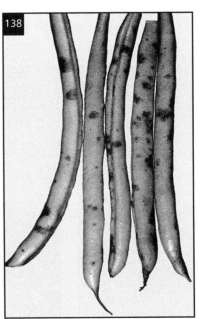

138 Green Phaseolus bean pod lesions caused by common blight (*X. campestris*).

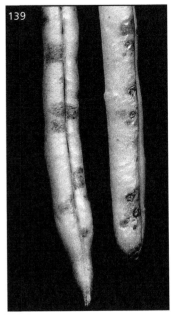

139 Grease spotting and necrosis caused by *X. campestris* to green Phaseolus bean pods.

SECTION 5

Viral Diseases

- ALFALFA MOSAIC VIRUS (AMV)
- BROAD BEAN STAIN VIRUS (BBSV) AND BROAD BEAN TRUE MOSAIC VIRUS (BBTMV)
- BEAN COMMON MOSAIC VIRUS (BCMV)
- BEAN CURLY TOP VIRUS (BCTV)
- BEAN LEAF ROLL VIRUS (BLRV)
- BEAN YELLOW MOSAIC VIRUS (BYMV)
- CUCUMBER MOSAIC VIRUS (CMV)
- PEA EARLY BROWNING VIRUS (PEBV)

- PEA ENATION MOSAIC VIRUS (PEMV)
- PEA SEED-BORNE MOSAIC VIRUS (PSBMV)
- PEA STREAK VIRUS (ALFALFA MOSAIC/RED CLOVER VEIN VIRUS) (PSV)
- PEA TOP YELLOWS VIRUS (PTYV) SYN. PEA LEAF ROLL VIRUS, BEAN LEAF ROLL VIRUS (BLRV)

ALFALFA MOSAIC VIRUS (AMV)

Host Crops
Green beans, dry beans.

Symptoms of Infection
Plants develop symptoms shortly after infestation by aphids. Plants may be stunted and develop distorted leaves and pods. There is often a bright yellow mottling or spotting of the leaves and some necrosis of the stem (**140, 141**).

Economic Importance
Occurrence of the disease in beans is relatively uncommon, although recent reports of more general infection have been made in Michigan, USA and, occasionally, in Idaho and Washington states. Yield losses will depend on the severity of infection. Where aphids are not controlled effectively early in growth, then some yield losses may be more significant.

Disease Cycle
The pea aphid (*Acyrthosiphon pisum*), bean aphid (*Aphis fabae*), and peach potato aphid (*Myzus persicae*) are commonly associated as vectors, although several other species have been recorded as vectors. The disease is seed-borne in alfalfa (lucerne) and aphids migrating from diseased alfalfa fields may transmit AMV to beans. It is not generally thought that AMV is seed transmitted in beans.

Prevention and Control
Effective control of aphids is essential where beans are grown in close proximity to alfalfa. There are a few AMV resistant varieties of Phaseolus beans available.

140 Leaf mosaic of alfalfa mosaic virus (AMV) on green Phaseolus bean foliage. (Courtesy of R. Larsen.)

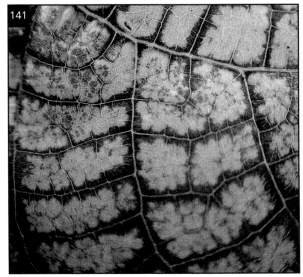

141 Leaf mosaic of AMV on green Phaseolus bean foliage. (Courtesy of R. Larsen.)

BROAD BEAN STAIN VIRUS (BBSV) AND BROAD BEAN TRUE MOSAIC VIRUS (BBTMV)

Host Crops
Field beans, broad beans.

Symptoms of Infection
Virus symptoms on beans are not always obvious. Both viruses may cause a range of slightly different symptoms on the leaves and pods. The viruses can produce a faint mottling with light and dark green areas visible on the leaves. Some leaf distortion can occur and, occasionally, plants from infected seed may be slightly stunted (**142**). The most obvious effect of BBSV is on the seed. Seeds from infected plants display a dark brown coloured area around the periphery (**143**). Pod fill may be reduced and the seed can be smaller than average.

Economic Importance
The effect on yield of either of the two viruses is generally small, although it may be directly proportional to the level of seed infection. The most severe effect is on quality where BBSV can result in bean seed spoilage, which is particularly damaging in broad beans grown for fresh market or processing. The presence of blemished beans may result in crop rejection by the processors.

Disease Cycle
Both viruses are seed-borne; they can then be transmitted from the infected seedlings to surrounding plants by weevil vectors. Mostly such transmission is by the clover seed weevil (*Apion vorax*). The bean weevil (*Sitona lineatus*) is a weak vector, although heavy and prolonged feeding by *Sitona* spp. can allow some virus spread to occur.

Prevention and Control
Because the viruses are primarily seed-borne, the use of healthy seed will prevent problems developing later on. Repeated use of home produced seed will facilitate the build-up of virus in the seed stock, so the use of new seed is recommended at regular intervals. The insect vector *Apion* spp. is very active and control is not practicable. Seed may be tested by specialist laboratories using an ELISA method.

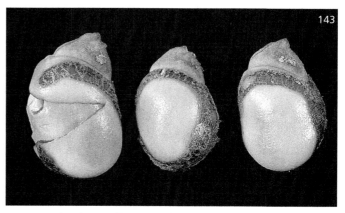

143 Necrotic edges on bean seeds caused by BBSV. (Courtesy of PGRO.)

142 Shoot distortion caused by broad bean stain virus (BBSV).

BEAN COMMON MOSAIC VIRUS (BCMV)

Host Crops
Green beans, dry beans.

Symptoms of Infection
Plants develop symptoms at any time from the production of the trifoliate leaves. The leaves show irregular-shaped pale and dark green areas over the surface. A mosaic pattern can develop but, more commonly, the leaves curl downward and may grow longer than others (**144**). Some beans may develop a brown discolouration on the leaf veins, stems, or pods and, occasionally, roots may be blackened.

Economic Importance
The virus can be found in all bean growing areas of the world. In some areas where susceptible varieties are grown, the disease can result in yield loss, poor pod set and pod development, and undersized seeds.

Disease Cycle
The virus can survive in weed hosts and, to a small extent, in infected seed. The transmission is then by aphid vectors including *Acyrthosiphon pisum*, *Macrosiphum euphorbiae*, *Myzus persicae*, and *Aphis fabae*.

Prevention and Control
Most commercial green bean varieties are resistant to BCMV, but dry bean types may be more susceptible. Efficient early aphid control and planting at a time before aphid activity occurs will also reduce the risk of infection.

144 Bean common mosaic virus (BCMV) symptoms of Phaseolus bean plant.

BEAN CURLY TOP VIRUS (BCTV)

Host Crops
Green beans, dry beans.

Symptoms of Infection
Seedlings develop as dwarfed plants, with severe puckering and downward rolling of the leaves. Plants are severely dwarfed and bunched. Leaves are brittle and the flowers may abort (**145, 146**). Any pods formed are stunted.

Economic Importance
The virus occurs primarily in the western United States and in British Columbia, Canada. Plant and yield loss can be high, especially in dry beans.

Disease Cycle
The virus has several hosts which are either perennial or winter annuals, such as Russian thistle and mustard and also sugar beet. It is transmitted by the beet leaf hopper, *Circulifer tenellus*. Warm dry conditions early in the season favour the migration of the leaf hopper from the overwintering hosts to newly emerging beans.

Prevention and Control
There are resistant varieties available. Insecticide treatment of the leaf hopper is usually too late to be effective.

145 Leaf distortion to green Phaseolus bean caused by bean curly top virus (BCTV). (Courtesy of R. Larsen.)

146 Chlorosis and death caused by BCTV. (Courtesy of R. Larsen.)

BEAN LEAF ROLL VIRUS (BLRV)

See also pea top yellows virus (PTYV)

Host Crops
Field beans, broad beans, peas.

Symptoms of Infection
Both autumn and spring planted beans can develop symptoms, usually during flowering and early pod set. Later planted or late maturing crops are more susceptible to BLRV as the virus vector, the pea aphid (*Acyrthosiphon pisum*), becomes more numerous in mid-summer. The upper leaves are tightly rolled; individual or small groups of plants may exhibit foliar symptoms of interveinal chlorosis (**147–149**). Pod set is reduced and those formed are often small and incompletely filled. Infected plants also become more susceptible to chocolate spot.

Economic Importance
Small groups of infected plants within a crop do not significantly affect yield, although the variability of seed size and poor seed set within the pods can reduce the value of beans for processing or fresh market. Where infection is more general within the field, then yield reduction is experienced.

Disease Cycle
BLRV is a persistent virus and can survive as a latent infection in a range of wild legume hosts, as well as in peas and Vicia beans. Aphids leaving their overwintering sites transmit the virus during periods of feeding on the host crop in the late spring and early summer. BLRV is not seed-borne.

Prevention and Control
BLRV is transmitted by the pea aphid. Usually infection is more obvious where aphids infest the crop before flowering. Control of early invading aphids is essential to prevent virus transmission, and recent work in spring sown crops indicated that a level of 5% of plants with pea aphids present represented a threshold for treatment with an aphicide.

147 Bean leaf roll virus (BLRV) on field bean plant.

148 Leaf roll and chlorosis caused by BLRV on field bean plant.

149 Interveinal chlorosis caused by BLRV on field bean plant.

BEAN YELLOW MOSAIC VIRUS (BYMV)

Host Crops
Field beans, broad beans, green beans, peas.

Symptoms of Infection
BYMV develops in beans at any time before flowering. Leaves are crinkled and may become pointed (150). Vein clearing can develop and the plant is slightly stunted. In field and broad beans, infection by pea enation mosaic virus (PEMV) is similar although the leaf crinkling is often accompanied by small translucent spotting and streaking on the leaf surface. Pods are often poorly developed, distorted, and with an uneven bumpy surface. In green beans and dry beans, leaves may droop and yellow mottling can occur on older leaves (151, 152). The plants may be dwarfed and stunted.

Economic Importance
The virus is widely distributed throughout the world and has several leguminous and nonleguminous hosts. The viruses are aphid transmitted and are common, especially in broad and field beans.

Infection may be on individuals or groups of plants. Depending on severity, the yield can be severely reduced and the produce blemished or uneven in size.

Disease Cycle
BYMV is also known as bean virus 2 or Phaseolus virus 2 in green and dry beans. It can be seed-borne in some species, including *Vicia faba*, but several aphid vectors can transmit the virus to beans from overwintering hosts. The pea aphid (*Acyrthosiphon pisum*) is the principal vector of both BYMV and PEMV, although black bean aphid (*Aphis fabae*) and peach potato aphid (*Myzus persicae*) can also be vectors. There is some evidence in the UK that persistent feeding by the bean leaf weevil (*Sitona lineatus*) may also result in cross-infection between plants of broad and field beans. BYMV and PEMV are persistent viruses and have a host range which includes peas, clover, lucerne, gladiolus, freesia, lupin, soya beans, and other wild legume hosts.

Prevention and Control
There are some resistant varieties of green and dry beans available, but early control of aphids in flowering bean crops is the most effective means of preventing wide-scale infection.

150 Bean yellow mosaic virus (BYMV) on field bean plant.

151 BYMV on Phaseolus bean leaves.

152 Necrotic mottling to Phaseolus bean plant leaves caused by BYMV.

CUCUMBER MOSAIC VIRUS (CMV)

Host Crops
Green beans, dry beans.

Symptoms of Infection
Symptoms can occur from the time of the first trifoliate leaf developing. Leaves are narrowed and pointed with a mosaic developing later on. Occasionally, early mosaic affected leaves may show some recovery. Foliar symptoms may include leaf curling, green or chlorotic mottling, and dark green vein banding particularly along the main leaf veins (153, 154). Leaves also develop dark green blisters and pods are small and misshapen, mottled, and curled (155). Where infection occurs late in growth or after flowering, leaf symptoms may not develop although pod effects may develop.

Economic Importance
The disease occurs in many countries including in Europe and the far East. In the USA, entire bean fields have been known to be affected but economic losses vary depending on the time of infection. However, where late infection occurs, the pod defects may result in the crop being unsuitable for processing.

Disease Cycle
Some strains of CMV can be seed-borne in Phaseolus, but the disease is readily transmitted by several aphid vectors. The virus has a wide range of weed hosts including several perennial species.

Prevention and Control
Healthy seed is a useful means of prevention but, where Phaseolus beans are planted regularly and in close proximity, then the aphid vectors will readily transmit the virus to healthy stocks. Management of aphids is important where the disease is a localized problem. Some commercial varieties are tolerant to CMV, and progress is being made to breed further for resistance.

153 Leaf doming on Phaseolus bean leaf caused by cucumber mosaic virus (CMV). (Courtesy of R. Larsen.)

154 Leaf distortion and blistering on Phaseolus bean leaf caused by CMV. (Courtesy of R. Larsen.)

155 Pod curling on Phaseolus bean caused by CMV. (Courtesy of R. Larsen.)

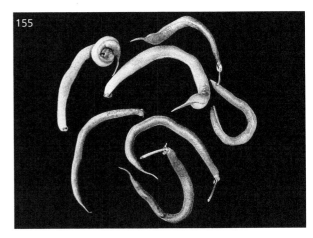

PEA EARLY BROWNING VIRUS (PEBV)

Host Crops
Peas.

Symptoms of Infection
The effects of PEBV may become obvious before flowering commences. Plants in defined areas are stunted (156), the stems are brittle, and the epidermis may split. A purple-brown discolouration develops on the stem towards the top of the plant. In other instances, symptoms may not become obvious until flowering. The upper leaves can display a distinct mottle with necrotic streaks between the veins (157, 158). The internal tissues of the stem are red-brown at the nodes. The top of the plant may be stunted and any pods are small. Occasionally, the pods produce clearly defined brown or purple coloured ring spots along the flat surfaces (159).

Economic Importance
PEBV has been recorded only in western Europe, and mainly in England and Holland. It is unknown in the USA and New Zealand. It is transmitted by free-living nematodes which are exclusive to sandy soil types. If the disease develops early on, the loss of yield from the stunted areas can be total. Where symptoms appear later, the maturing of the crop can be affected within the infected areas, which can cause a problem in vining peas where the harvested crop contains peas of mixed maturities. It is not uncommon to find the disease in crops growing in fields with no previous history of pea growing.

Disease Cycle
PEBV is a tobra virus and is transmitted by the stubby root nematode, *Trichodorus* spp., which survives in a free-living form in sandy soils. In the spring, when soil conditions are moist, the

156 Stunted pea crop caused by pea early browning virus (PEBV). (Courtesy of PGRO.)

157 Leaf mosaic caused by PEBV. (Courtesy of PGRO.)

158 Terminal shoot necrosis caused by PEBV.

159 Ring spotting on pea pods caused PEBV. (Courtesy of PGRO.)

nematodes graze on the newly developed roots of the pea seedlings (160) and the virus is passed to the plants. As the soil dries, the nematodes migrate down the soil profile and remain in moist soil for the rest of the year. The virus can infect a range of weeds as well as *Vicia faba* beans and sugar beet, although no symptoms are seen on these alternative hosts. Infected peas can transmit the virus to the seed and this could be an important means of infecting otherwise clean soil. However, the nematodes are an essential means of virus transmission to peas and in the absence of nematodes, the disease does not survive.

Prevention and Control

Peas growing on sandy soils in Europe, where *Trichodorus* spp. are known to exist and sugar beet is grown regularly, may be at risk if the soil remains wet after drilling. However, the problem is not common every year and only those fields where the disease is seen regularly should be avoided for future pea production. The risk of seed transmission is only very slight and where seed from infected crops is grown in soils other than sandy, no symptoms will develop. There is no effective means of seed testing and no varieties have been shown to be resistant. Nematicides such as metam sodium or aldicarb may be useful, but the cost of treatment is high.

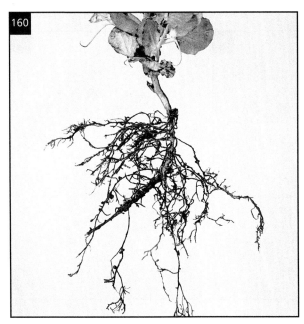

160 Free-living nematode damage to pea roots.

PEA ENATION MOSAIC VIRUS (PEMV)

Host Crops
Peas, field beans, broad beans.

Symptoms of Infection
After flowering, peas in small to large patches distributed over the field develop a mosaic and mottled leaf symptom at the top of the plant (161–163). Vein clearing can occur and the veins may also be ridged. The newly developing leaves are often small and distorted. Older leaves may also develop mottles, small irregularly-shaped translucent flecks and, on occasion, thin flaps of tissue or enations may develop along the leaf veins, hence the name of the virus (164). In severe infections, a proliferation of adventitious flowering shoots develops from the leaf axils. These do not produce pods and the flower petals may remain a green colour. Pods which do form from the main stem are roughly ridged, covered with raised lumps (165). Peas inside can be malformed or underdeveloped.

Economic Importance
In Europe, the disease can be common in most years, as it is transmitted by the pea aphid (*Acyrthosiphon pisum*) and peach potato aphid (*Myzus persicae*). PEMV also occurs in China, Iran, and Canada. In the USA, infections may not be so common, as many varieties have been selected for resistance. However, the disease is indicative of earlier aphid infestation and the combination of insect damage and virus can reduce yield overall.

Disease Cycle
PEMV has an overwintering host range which includes wild legumes as well as cultivated legumes such as *Vicia faba* (166, 167), chickpea (*Cicer arietinum*), sweet pea (*Lathyrus odoratus*), lentil (*Lens culinaris*), *Medicago arabica*, *Trifolium incarnatum*, and *Vicia sativa*.

Prevention and Control
Several varieties are resistant or tolerant to the virus. However, early aphid control at the beginning of flowering will reduce the severity of infection.

161 Mottling and chlorosis of pea foliage caused by pea enation mosaic virus (PEMV).

162 Interveinal mottling and chlorosis of pea leaves caused by PEMV.

163 Translucent mottling to pea foliage caused by PEMV.

164 Leaf enation caused by PEMV. (Courtesy of PGRO.)

165 Pod blistering on peas caused by PEMV.

166 Mottling and chlorosis of field bean (*Vicia faba*) caused by PEMV.

167 Pod distortion of field bean (*V. faba*) caused by PEMV.

PEA SEED-BORNE MOSAIC VIRUS (PSBMV)

Host Crops
Peas, broad beans, field beans.

Symptoms of Infection
Peas become infected at any time during the growing season, but if the source of the infection is from the seed, the seedlings are stunted from the onset of growth (**168**). Individual plants scattered randomly throughout the crop are indicative of a seed-borne infection source. These plants may develop rolled leaves, be pale in colour, and any pods formed are small, undersized, and contain few seeds (**169**). The virus is also aphid transmitted. Where plants acquire the virus before flowering, the effects are more obvious as larger areas of the crop show symptoms. Often the infection is by both PSbMV and pea enation mosaic virus (PEMV). Where this occurs, the plants are dwarfed, leaves rolled, and there is a proliferation of adventitious shoots from the leaf nodes which, in turn, produce very small unproductive flower buds. Seeds are undersized or the testa can display a blistering or blemish which can resemble the markings on a tennis ball (**170, 171**). Not all varieties of peas express the same symptoms.

Economic Importance
PSbMV is probably the most serious virus disease of peas. Because it is seed-borne, it has been recorded in all the pea growing countries of the world. It has also been spread inadvertently in pea germplasm, and has resulted in many breeding lines carrying the virus. Yield loss can be severe and in the UK in the mid- to late 1980s it was responsible for the loss of hundreds of tonnes of peas of a single variety produced for quick freezing. Because broad beans and field beans are also hosts to the virus, the siting of peas near to these crops should be avoided.

Disease Cycle
Once the virus has been acquired by the plant, the virus multiplies within the cells and spreads throughout the actively growing tissues. The virus enters the developing seed via the peduncle and, once in the testa, is able to move into the embryo. Infected embryos are the principal means of seed transmission. Aphid species including the pea aphid (*Acyrthosiphon pisum*) and the peach potato aphid (*Myzus persicae*) can transmit the virus. Cereal aphids have also been implicated as PSbMV is a nonpersistent virus and is therefore easily spread by migrating aphids as they probe for a suitable host.

Prevention and Control
Aphid control is of limited value as the aphids often pass the virus to surrounding plants before they are affected by any contact acting or systemic insecticide. The virus will also infect *Vicia faba* and this crop can also be a source of infection if growing in close proximity to peas. The use of healthy seed stocks is the only reliable means of prevention. Seeds or seedlings can be tested for virus presence and only those stocks where the virus is absent should be used both for commodity crops and for seed production. An increasing number of varieties is being developed with resistance to PSbMV and the wider adoption of these will substantially reduce the risk of disease in the future.

168 Stunting and dwarfing caused by pea seed-borne mosaic virus (PSbMV).

169 Poor pod development caused by PSbMV.

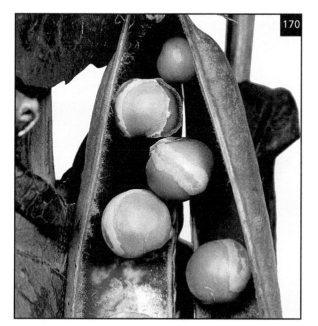

170 'Tennis ball' blisters on peas inside the pod caused by PSbMV.

171 Irregular sizes of peas after crop infection with PSbMV.

PEA STREAK VIRUS (ALFALFA MOSAIC/RED CLOVER VEIN VIRUS) (PSV)

Host Crops
Peas.

Symptoms of Infection
There can be several viruses in the PSV complex, which may include red clover vein virus and alfalfa mosaic virus, although PSV can occur on its own. Symptoms generally appear later on in the growing season when the pods are formed but not filled. The top of the plant may show a distinct mottle and chlorosis and the leaves appear brittle. There may be several other spots and streaks on the leaves, and there may be some discolouration of the vascular tissue in the stems (172). Pods may remain unfilled, the surfaces display numerous small pit marks, and there can be a general purple discolouration of the pod (173, 174). Where infection occurs early on, the plants can become severely affected and the crop can die before flowering has begun.

Economic Importance
The virus can be found wherever pea aphids (*Acyrthosiphon pisum*) are present as these are the prime vectors. However, in Europe, the virus appears sporadically and seems to be more common during hot dry summers; otherwise the principal virus found in peas is pea enation mosaic virus. PSV is more common in the western USA where alfalfa is considered to be the major overwintering host of both the virus and the aphids.

Disease Cycle
The overwintering hosts of PSV are lucerne (alfalfa) and red clover. Other hosts are known, including *Vicia faba* and soya bean, but they may be symptomless. The virus is carried and transmitted in a nonpersistent way by the pea aphid, and is acquired after a very short period of feeding. In cool summers, the aphid feeding may be delayed during periods of unfavourable weather and, therefore, there is less risk of PSV surviving in the vector.

Prevention and Control
USA pea breeders are currently screening varieties for tolerance. There is a small number of commercial varieties with good field tolerance to PSV and also pea top yellows virus (PTYV). Effective aphid control is essential early in the season, to reduce the risk of large numbers of the pea plants becoming infected.

172 Pea streak virus (PSV). (Courtesy of PGRO.)

173 Purple pod symptom of PSV.

174 Poor pod fill caused by PSV.

PEA TOP YELLOWS VIRUS (PTYV) SYN. PEA LEAF ROLL VIRUS, BEAN LEAF ROLL VIRUS (BLRV)

Host Crops
Peas, field beans, broad beans.

Symptoms of Infection
The most obvious symptom is a yellowing of the upper part of the plant. Upper leaves may be slightly upwardly rolled and pods may be undersized and more pointed at the end (175). Infected plants occur in patches or larger areas over the field, and these correspond precisely with early pea aphid infestation. The plants develop symptoms usually after the onset of flowering.

Economic Importance
Although the disease is also found extensively in *Vicia faba* beans, peas seldom show severe symptoms. Many varieties appear to be resistant or tolerant to PTYV, although some varieties may be symptomless hosts. Only where the disease is widespread, or where there is infection by other aphid transmitted viruses, may there be a significant reduction of yield.

Disease Cycle
The virus is present in a range of wild hosts such as clover or vetches and in lucerne (alfalfa) and winter sown field beans (*Vicia faba*). The principal vector is the pea aphid (*Acyrthosiphon pisum*), which can overwinter on infected hosts. As the virus is carried in a persistent manner in spring, the aphids migrate to legume crops and the virus is transmitted to the summer host crop. Aphid populations move back to the overwintering hosts shortly before harvest.

Prevention and Control
Many pea and bean varieties are resistant or tolerant and these should be used, particularly in later plantings which may be more susceptible to aphid infestation. Where aphids are detected early in the season, just as flowering commences, effective treatment should be applied to reduce the risk of a severe virus infection.

175 Pea top yellows virus (PTYV).

SECTION 6

Pests of Stem, Foliage, and Produce

- PEA APHID
- CUTWORMS
- BLACK BEAN APHID
- SILVER Y MOTH
- PEA SEED BEETLE
- BEAN SEED BEETLE
- TORTRIX MOTH
- PEA MIDGE OR PEA GALL MIDGE
- PEA MOTH
- SLUGS AND SNAILS

- STEM NEMATODE
- HELIOTHIS CATERPILLAR (CORN EARWORM)
- PEA CYST NEMATODE
- PEA THRIPS
- LEAF MINERS
- ROOT KNOT NEMATODE
- TWO-SPOTTED SPIDER MITE
- STUBBY ROOT NEMATODE

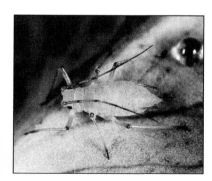

PEA APHID:
Acyrthosiphon pisum

Host Crops
Peas, field beans, broad beans.

Symptoms of Infestation
Aphids are likely to infest peas at any stage after crop emergence, although in early planted combining peas, infestations are seldom serious before flowering has commenced. Yield loss is caused by the direct feeding effect of aphids. In addition, however, several aphid transmitted viruses (pea enation mosaic, pea seed-borne mosaic, and bean leaf roll or pea top yellows virus) are associated with infested crops, particularly if aphids are present before the onset of flowering.

Damage is characterized by the presence of smothering colonies of the large green aphid which can result in growth distortion of the growing point, loss of flowers, and undersized distorted pods which often fail to fill (**176–178**). During feeding the aphids produce honeydew that is an ideal medium for colonization by saprophytic fungi, and *Botrytis cinerea* infection can occur if the weather conditions are wet and favourable for the fungus (**179**).

Aphids also attract predators particularly hoverflies (Syrphidae), the pupae of which may become serious contaminants during the harvest of vining peas (**180**).

Economic Importance
Pea aphid is a major pest in all temperate areas of the world where peas or other legumes are grown. Aphids can cause serious losses to peas when populations colonize the growing points. Yield losses of up to 45% are reported to have occurred in peas for freezing or canning; in dry harvest peas, losses of around 20–30% have been experienced.

Pest Cycle
The pea aphid is bright green in colour, with a pear-shaped body, slightly dark legs, and red eyes (**181**, **182**). Both winged (alate) and wingless (apterous) forms are found within the colony. Nymphs are produced viviparously and the final instars can themselves reproduce after about 6 days. Aphids can overwinter on wild host crops such as vetches, clovers and, sometimes, lucerne (alfalfa) as eggs or as adults. The winged generation produced on the winter host migrates to peas in the spring. At the end of the season when the summer hosts are near to harvest, a winged generation returns to the overwintering host plants where males and females are produced which, in turn, produce eggs.

Prevention and Control
Control of aphids is best achieved with systemic insecticides such as pirimicarb. Treatment should be applied as soon as colonies can be found on 15–20% of the plants, particularly if the crop is at the early flowering stage. Where aphid infestation develops later in the growing season, yield increases from insecticides can be obtained in combining peas up to the time of the development of the fourth pod-bearing node, but in vining peas, such increases can only usually be obtained by the first pod fill stage. An aphid population prediction model is available in the UK (PAM) which uses air temperatures to predict the time at which an economic threshold is reached before insecticides need to be applied.

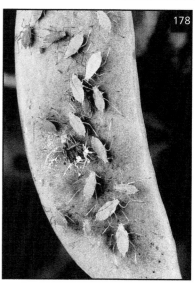

176 Pea aphid (*Acyrthosiphon pisum*) infestation on pea plant.

177 Pea aphid (*A. pisum*) colony on young pea pod.

178 Pea aphids (*A. pisum*) some with fungal infection *Pandora neoaphidis*.

179 Necrosis after pea aphid (*A. pisum*) infestation in pea crop.

180 Hoverfly pupae (*Syrphidus* sp.) crop contamination following aphid infestation. (Courtesy of PGRO.)

181 Pea aphid (*A. pisum*).

182 Pea aphid (*A. pisum*) feeding.

CUTWORMS:
Agrotis segetum

Host Crops
Green beans.

Symptoms of Infestation
Caterpillars feed just below the surface of the soil in dry periods during early to mid-summer. Seedlings are severed and the stems are left lying on the surface. Often the damage occurs along short runs of row.

Economic Importance
Severe plant loss affects yield and a crop with gaps allows weeds to develop uncompetitively.

Pest Cycle
Cutworms are the larvae of a group of Noctuid moths; the turnip moth is the most common. The moth is widely distributed throughout temperate regions but not in the USA or Australasia. The turnip moth has a wingspan of 40 mm and is grey-brown in colour, with darker markings of rings and lines. The hindwings are lighter in colour (183). Eggs are laid during early summer on soil or plant debris. After about 21 days the larvae hatch and begin feeding on the foliage. As they increase in size, the larvae feed on the soil surface attacking young plant stems (184). Larvae are grey to brown in colour but are sometimes tinged with green. They have a dark line along the back and lighter lines along the sides. The larvae reach 40 mm in length when mature (185).

Prevention and Control
The caterpillars are easily controlled with pyrethroid sprays, but usually damage has occurred before treatment can be made. Irrigation will reduce larval activity.

183 Turnip cutworm moth (*Agrotis segetum*).

184 Cutworm (*Agrotis* sp.) damage to young pea plant.

185 Turnip cutworm (*A. segetum*) caterpillar.

BLACK BEAN APHID:
Aphis fabae

Host Crops
Field beans, broad beans, green beans, runner beans, peas.

Symptoms of Infestation
Colonies of black bean aphid develop rapidly on the upper parts of Vicia beans during early summer. As the colonies develop, the aphids move further down the stems and colonize the developing pods. At first, individual stems may be infested; later, the aphids spread to surrounding plants, developing into localized patches of infested crop (**186–188**). The aphids are black, 1–2 mm in size, with small grey-white speckles on the dorsal surfaces (**189**, **190**). Infested plants fail to fully develop pods, and if colonized early in the season, may produce symptoms of leaf rolling due to infection by bean leaf roll virus. In green and runner beans, populations remain small and discrete colonies develop on the flowering shoots and pods.

186 Black bean aphid (*Aphis fabae*) colony on Vicia bean plant.

187 A group of mainly alate with some apterous black bean aphids (*A. fabae*) on Vicia bean plant.

188 Black bean aphid (*A. fabae*) colony on Vicia bean plant affecting flower development.

189 Reproducing colony of black bean aphids (*A. fabae*).

190 Photomicrograph of black bean aphids (*A. fabae*) feeding alongside leaf vein.

Economic Importance

Early infestation of beans results in the largest effect on yield. Pod numbers are reduced and seed development may be affected. Where pod infestation occurs, the produce may be unsaleable as fresh picked pods. The sticky honeydew produced during feeding can become colonized by saprophytic moulds which, in turn, further reduces the value of the fresh market pods.

Pest Cycle

In Europe, the aphids overwinter as eggs on woody host plants, especially the European spindle (*Euonymus europaeus*) (**191**). In early spring, the eggs hatch and apterous females are produced which, in turn, produce a generation of alate females which fly to the summer hosts (**192**). This generation produces colonies of viviparous alate females which form dense colonies. As the crop matures, a winged generation is produced which migrates to the winter hosts and give rise to male and female aphids. Eggs laid by the females overwinter on the stems of the shrubs.

Prevention and Control

Some weeds are also summer hosts to black bean aphids, and efficient control of thistle and fat hen (*Chenopodium album*) will reduce localized infestation. In small crop areas, the tops of the bean plants can be trimmed just after pod fill to reduce the availability of colonizing sites to migrating aphids. Insecticides may be applied to beans when aphids can be found on 10% of plants, especially as they reach the early flowering growth stage.

191 Black bean aphids (*A. fabae*) on foliage of spindle tree (*Euonymus* sp.), an overwintering host.

192 Black bean aphid (*A. fabae*) alates.

SILVER Y MOTH:
Autographa gamma

Host Crops
Peas, field beans, broad beans, green beans, runner beans.

Symptoms of Infestation
The main cause of damage by the caterpillar of the Silver Y moth is by feeding on the foliage or pods of peas or beans (**193**). In addition, contamination by the caterpillar or the pupae in peas for processing can result in crop rejection (**194, 195**). The moth is a frequent migrant to northern Europe and Scandinavia from north Africa, where it can develop in high populations before being pushed north by southerly winds during the early part of the summer.

Economic Importance
Feeding damage can be severe in years of high pest incidence, but foliar damage is rarely the problem in peas or beans. Pod damage to peas or beans grown for fresh picking can reduce the quality and hence the yield when damaged pods have to be removed before sale. Contamination of vined peas by the caterpillar or the pupae can result in crop rejection by the processors if detected in the pre-load inspection samples; if undetected, contamination can be a major cause of customer complaints to the retailer.

Pest Cycle
The silver Y moth is a large, day flying Noctuid moth with a wingspan of around 35–40 mm. The wings are grey-brown with pale markings and a distinct silver 'Y' in the middle of each forewing. The body is hairy and thickened (**196**). Adults

193 Silver Y moth (*Autographa gamma*) caterpillar.

194 Silver Y moth (*A. gamma*) caterpillars contaminating vined peas. (Courtesy of PGRO.)

195 Silver Y moth (*A. gamma*) pupae contaminating vined peas. (Courtesy of PGRO.)

196 Silver Y moth (*A. gamma*).

PESTS OF STEM, FOLIAGE, AND PRODUCE

197 Silver Y moth (*A. gamma*) pupating in webs around pea leaves. (Courtesy of PGRO.)

migrate during early to mid-summer and are attracted to a wide range of flowering plants and leafy vegetables. Round white eggs are laid singly on the foliage, and larvae hatch after 10–14 days. They begin feeding on the leaves first, causing a circular crater where the leaf surface is eaten; as the caterpillar becomes larger, then parts or whole leaves and parts of pods are devoured. The caterpillars are bright green with a white line along each side of the body and a darker line along the back. There are only three pairs of abdominal prolegs. The head is a darker brown-green, and the caterpillar moves over the plant with a looping action. When disturbed, the caterpillar rolls into a ball. When mature after 3 weeks, the caterpillar spins a web in the foliage (**197**) and produces a black shiny chrysalis. After a further 10 days, the newly emerging adults fly to other hosts but, rarely, produce a second brood of caterpillars. In the UK, local populations of adults can overwinter.

Prevention and Control

Control of caterpillars with pyrethroid insecticides is easily achieved when they are present in the crop. A monitoring system consisting of traps containing the silver Y female sex pheromone is available to detect large numbers of migrating adults. In the UK, the system has been developed to define treatment thresholds in peas for vining to avoid losses caused by caterpillar contamination.

PEA SEED BEETLE:
Bruchus pisorum

Host Crops
Peas.

Symptoms of Infestation
The first sign of injury to peas is when the adult beetles emerge from the mature seed, leaving a circular exit hole. Emergence can occur in the field just prior to harvest of the dried crop or shortly afterwards, while the seed is being stored. The beetles leave the seed and migrate to surrounding shrubby vegetation where they overwinter.

Economic Importance
The pea seed beetle is found in the USA and in many of the warmer European countries, but not in the UK or Scandinavia. Peas grown for dry harvest and destined for the premium human consumption or seed markets are blemished by the holes. Locally high populations of Bruchids can develop and, once established in an area, the infestation level can be too high for the pea or bean processors to clean and sort. This devalues the crop; where mechanical sorting is possible, such cleaning incurs price penalties to the producer. Damaged seed may still germinate; however, the smaller sized pea seed may succumb to pre-emergence mortality in field conditions and the presence of damaged peas in seed is considered undesirable by the end user.

Pest Cycle
Adult beetles leave their overwintering sites in early summer, when maximum temperatures reach 20°C (**198, 199**). They readily fly to flowering crops and seem to be attracted by the flower volatiles. After feeding, eggs are laid singly on newly developing pods. The eggs are 0.3–0.5 mm in length, elongated, and semi-transparent. The basal edge is cemented onto the pod surface. After a few days, the eggs hatch and the larvae bore through the bottom of the egg case and into the pod wall. They penetrate the immature seed and begin feeding on the cotyledons below the testa. Feeding continues until the seed is

mature, when the larvae pupate and cut their way out of the dry seed coat and leave the seed (**200**).

Prevention and Control

Control is difficult to achieve as the beetles feed within the crop canopy and are not often exposed to contact acting insecticides. There are few insecticides with sufficient ovicidal activity to reduce damage once the eggs are laid. Early detection of adults in flowering crops is essential so that spraying can take place before egg laying begins. A second application may be necessary to protect the later developing pods. Early flowering pea varieties may escape damage if the adult flights from the over-wintering sites are delayed by cool temperatures.

198 Pea seed beetle (*Bruchus pisorum*).

199 Pea seed beetle (*B. pisorum*).

200 Peas infested with pea seed beetles (*B. pisorum*).

BEAN SEED BEETLE:
Bruchus rufimanus

Host Crops
Field beans, broad beans.

Symptoms of Infestation
Beans are susceptible to infestation as they begin flowering. Eggs of the bean seed beetle are 0.5–1 mm long, cigar-shaped, a pale cream to yellow colour, and are laid on the pod surface. After hatching, the larvae tunnel through the pod wall into the developing seed. As they mature, the larvae are white, segmented, legless 'grubs' with a brown head. They feed within the cotyledon of the seed (**201**). As the seed matures, the larvae pupate and the new generation of adults emerges from infested seed by biting away the seed coat covering the cell in which it has pupated (**202**). The adults are oval beetles 2–3 mm long, with black wing cases which do not completely extend over the abdomen. Small white markings are visible on the slightly rough surface of the wing cases (**203, 204**). Infested immature seed exhibits small cuts in the seed coat surface where the larva has entered (**205**). Mature seed shows circular windows covering the holes in which the pupae develop. Emerging adults leave a circular hole in the seed.

Economic Importance
Fresh harvested broad beans may be blemished by larval entry holes and contain developing larvae in the cotyledon. Infested produce for processing results in crop rejection by the processing companies. In dry seed, the presence of pupae or unemerged adults reduces the value of the produce for seed or for use in human consumption.

Pest Cycle
Bruchus rufimanus occurs throughout Europe. The beetle produces one generation each year. Adults migrate to flowering beans in early summer when temperatures reach daily maxima of 17–20°C. They feed on pollen until sexually mature, after which time eggs are laid on the smallest pods at the lower parts of the stems. The greatest concentration of damage occurs to seeds produced by the lowest three or four reproductive nodes. After pupating in the seed, the adults emerge, either in the field before harvest, or later in storage. A parasitic wasp (*Triaspis luteipes*) can sometimes be found emerging from the seed after pupation. This leaves a smaller round hole in the seed coat. Adult Bruchids migrate to overwintering sites, such as field edges and woody vegetation. A closely related species, *Acanthoscelides obtectus*, attacks Phaseolus beans and produces several generations per year as it breeds in stored produce (**206**).

Prevention and Control
Both winter and spring sown field and broad bean crops are susceptible to damage by *B. rufimanus*, and the commonly grown European varieties show no evidence of varietal resistance. It has proved to be very difficult to control the pest in crops. Once eggs are laid, the larvae are unexposed and control by insecticides is not possible. Control of adults during the pre-egg laying feeding period has met with limited success, but correct timing of sprays is essential to prevent egg laying. This is best achieved when 5–6 flower trusses have developed on the plants and no sign of pod formation is visible. A second spray application, 7–10 days later is recommended where infestation is prolonged.

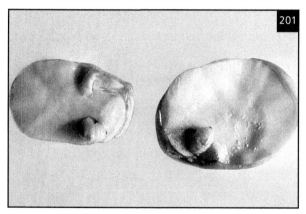

201 Bean seed beetle (*Bruchus rufimanus*) larvae in Vicia bean seed. (Courtesy of PGRO.)

202 Bruchid larvae (*B. rufimanus*) in Vicia bean seeds. (Courtesy of PGRO.)

203 Bean seed beetle (*B. rufimanus*).

204 Bean seed beetle (*B. rufimanus*).

205 Bean seed beetle (*B. rufimanus*) larval entrance hole in Vicia bean seed.

206 Seed beetle (*Acanthoscelides obtectus*) infestation in Phaseolus bean seeds. (Courtesy of Holt Studios/Duncan Smith.)

TORTRIX MOTH:
Cnephasia asseclana

207 Leaf webbing in peas caused by tortrix larvae (*Cnephasia asseclana*). (Courtesy of PGRO.)

Host Crops
Peas.

Symptoms of Infestation
The damage caused by the larvae of the tortrix moth is observed in the spring just before flowering commences. The upper leaves are tied together with a fine web, giving the plants a hooded appearance (**207**). Within the bunched leaves, the caterpillar can be found feeding on the leaves of the growing point. The moth is found in most parts of Europe and infests a wide range of plants, both crops and flowers.

Economic Importance
Damage is very seldom severe enough to cause any loss of plants or yield. The caterpillars have usually pupated and the new generation of moths have left the crop before harvest.

Pest Cycle
The moths are a dull greyish brown with a wing span of around 16 mm. The moth is largely nocturnal. Eggs are laid singly on the upper leaves of the growing point. After 7–10 days, the caterpillar begins to feed on the foliage, producing a web as it does so, which pulls together the leaves providing a shelter (**208, 209**). After 2 weeks, the larvae pupate and the moths fly to other hosts.

Prevention and Control
No control measures are justified.

208 Tortrix (*C. asseclana*) caterpillar feeding on Vicia bean leaf.

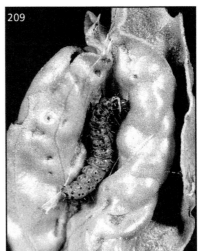

209 Tortrix (*C. asseclana*) caterpillar beginning to pupate in Vicia bean leaves.

PEA MIDGE OR PEA GALL MIDGE:
Contarinia pisi

Host Crops
Peas.

Symptoms of Infestation
This tiny Cecidomyid midge can result in loss of yield in peas in areas of locally intensive production. It is present in the cooler western temperate areas of the world but is a particular problem in the UK, France, Holland, Germany, Switzerland, Austria, Scandinavia, and Russia. The damage to the crop is caused by larval feeding within the developing flower bud, during the summer. This often results in a distortion of the bud, and a nettle head caused by a foreshortening of the flower stalks. Damaged buds fail to produce pods (**210**).

The damaged buds contain numerous small white segmented larvae about 1–2 mm in length (**211, 212**). When mature, the larvae flex themselves in a rapid movement to throw themselves out of the plant where they wriggle down into the soil. The adult midge is grey-brown in colour, with a body length of 2 mm. There is one pair of very fine long wings and the legs are slender and longer than the body. The midge can be found resting on the developing flower buds (**213, 214**).

210 Pea midge (*Contarinia pisi*) damage to pea shoot.

211 Pea midge (*C. pisi*) larvae inside flower bud.

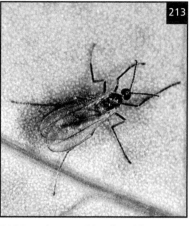

212 Pea midge (*C. pisi*) larvae.

213 Female pea midge *(C. pisi)* at rest.

214 Male pea midge (*C. pisi*).

215 Pea midge (*C. pisi*) larval damage to Vicia bean pods.

Economic Importance

Vining peas are often severely damaged by pea midge. Because pea varieties have been developed to produce a short flowering period to allow more even pod maturity, a large proportion of flower buds is susceptible to attack by the midge larvae. Combining peas, which are less determinate and have a longer flowering and pod setting period, suffer less damage as there are fewer flower buds available to attack during the relatively short period of midge activity. Where there are a larger number of flower buds within the apical shoot damaged by larvae, yield loss can be as much as 50%.

Pest Cycle

There is usually a single generation each year. After pupation in the spring, the adults emerge from the previous year's pea field and, after mating *in situ*, the females fly to nearby pea crops during the latter part of the day. The females lay eggs on the developing flower buds of crops which are still enclosed by the terminal leaves. After about 4–5 days, the eggs hatch and the larvae burrow into the bud where they feed inside the base. After another 5–7 days the larvae leave the buds and fall to the soil, where they develop a cocoon in which they begin diapause. In late spring when soil temperatures have increased to 15–17°C, the larvae pupate and adults emerge shortly afterwards. Midge larvae have occasionally been found infesting Vicia bean pods (**215**).

Midge tends to emerge over a very short period, often over 1–2 days. This makes control and detection very difficult. The prime source of infestation is from the previous year's pea fields and where the migration distance is short, then localized high populations can develop.

Prevention and Control

Susceptible crops are those which have reached the enclosed bud stage. In areas where midge was a problem the previous year, crops should be examined for midge adults by pinching together the outer leaves of the growing point and peeling back the leaves to reveal the buds. If adult midges can be found, an insecticide such as lambda-cyhalothrin should be applied as soon as possible, to reduce the risk of eggs being laid.

A pheromone-based monitoring system is currently available in the UK and France, to aid detection of emergence and identify susceptible crops. In some areas it may be possible to avoid serious damage by avoiding growing peas close to previously infested fields. It may also be possible under some circumstances to practice large-scale rotation, which includes neighbouring farms.

PEA MOTH:
Cydia nigricana

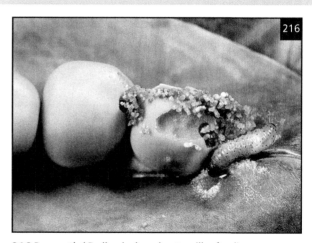

216 Pea moth (*Cydia nigricana*) caterpillar feeding inside pod.

Host Crops
Peas.

Symptoms of Infestation
Damage by the pea moth larvae occurs inside the pods, where feeding on the developing seeds results in irregular-shaped holes or notches. Also present in the pod is the frass, left during feeding (216). Pea moth is a common European and east European pest of peas. As yet, no reports of this species have been recorded in the USA.

Economic Importance
While yield loss is seldom severe, the produce of peas grown for human consumption or seed is blemished. In vining peas, there is a risk of contaminating the pack with the caterpillar, the blemished pea, or the frass and it is likely that peas containing such damage will be rejected by the processing factory. Peas harvested dry are the most affected because of the extended time of crop maturation in the field, which allows the completion of the larval development. Peas for the quality market are down-graded, and damaged seeds are classed as waste and are mechanically removed by the processor. This often results in payment deductions to the grower.

Because damage is largely cosmetic and yield loss is not generally serious, the moth is of little consequence to crops being produced for the animal feed compound market. However, in areas where moth populations are allowed to build-up unchecked, the pest becomes more of a problem to producers of high quality combining peas or peas for human consumption.

217 Pea moth (*C. nigricana*).

Pest Cycle
The adult moth is quite small with a wing span of 12–15 mm. It is silvery brown in colour, with indistinct black and white markings at the tips of the forewings (217). It flies during the sunny parts of the day, when moths can be seen flying close to the flowering pea crop. The larva, or caterpillar, is

218 Pea moth (*C. nigricana*) caterpillar feeding in pea pod.

creamy white with a brown to black head. The length of the final instar of the larva is about 6 mm. It is found inside pods either singly or in pairs (218).

The adult moths emerge from the soil of the previous year's pea crop and fly to flowering crops in early to mid-summer. The females lay small flattened eggs on the leaves, stems, or stipules of flowering plants (**219**, **220**). Eggs hatch in about 8–17 days depending on the temperature, and the larvae then move over the surface of the plant and bore into the young developing pod (**221**). Once inside, the larvae feed on the peas for around 3 weeks, after which time the pods of the dry harvest peas are mature. The fully developed larva bites its way through the pod wall and falls to the ground, where it buries itself in the soil. A silken cocoon is spun around the larva, and it remains in the soil in this state until the following summer. Pupation is triggered by rising temperatures and emergence occurs about 5 days later.

Prevention and Control

Moth attacks are most frequent and most damaging in intensive combining pea growing areas, where local high populations can develop. Control of the larvae has to be carried out before they reach the pod, after which time little effect is gained by insecticide sprays. Spraying has to be done at the point at which the eggs are hatching. In the UK this is achieved using a pheromone-based monitoring system. The pheromone is contained in a delta-shaped trap, and placed in the pea field just prior to the commencement of flowering (**12**). Male moths are attracted by the female sex pheromone, and are caught in the sticky traps just as the main migration of moths occurs from the overwintering sites. Depending on the number of moths caught in a specified period, a decision can be made as to the need for treatment. Egg development is directly related to air temperature, and a correlation between average day temperatures and egg hatch allows sprays to be timed accurately. Often a second application 10–14 days after the first will prevent damage from later arriving moths.

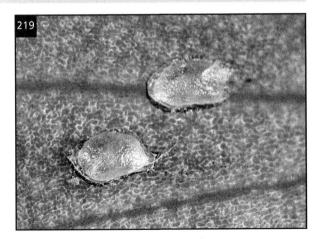

219 Newly laid pea moth (*C. nigricana*) eggs.

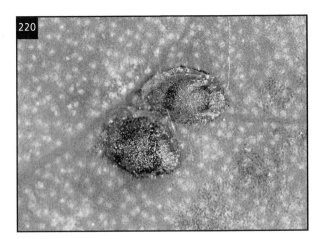

220 Pea moth (*C. nigricana*) eggs about to hatch.

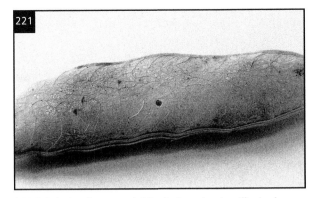

221 Exit hole of pea moth (*C. nigricana*) caterpillar in dry pea pod. (Courtesy of PGRO.)

SLUGS AND SNAILS:
Deroceras reticulatum,
Cernuella sp., and *Cepaea* sp.

Host Crops
Peas, field beans, broad beans, green beans, runner beans.

Symptoms of Infestation
Slugs are the most widespread invertebrate pest in horticultural and agricultural crops. They are more common in soils with a high clay content, and are prevalent where there is an abundance of undecomposed cereal straw residue or vegetable residues from the previous crop. Damage to seeds or young seedlings can occur from any time after planting. Slugs feed by rasping the plant tissue, leaving a ragged edge or severed shoot and leaves (**222**). The seed may be hollowed out and the developing seedlings fail to emerge from the soil (**223**, **224**). The pests are widely distributed in the field. In wet seasons, slugs also feed on the foliage during the day as well as at night and, if present during the harvest of vining peas, they become mixed with the vined peas and become contaminants (**225**).

222 Grey field slug (*Deroceras reticulatum*) on damaged pea plant.

223 Slug damage to pea seedling.

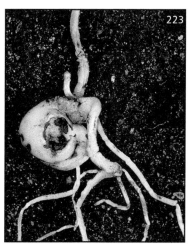

224 Slug damage to winter field beans. (Courtesy of PGRO.)

225 Slug contamination in vined pea crop. (Courtesy of PGRO.)

226 Banded snail (*Cepaea* sp.).

Snails live in field margins and are most commonly found in calcareous stony soils. They migrate into the crop during wet weather conditions, often using the lodged foliage to crawl (**226**). Snails are also liable to become a contaminant problem in vining pea crops.

Economic Importance
Loss of seed and seedlings may be high where populations of slugs are high, particularly in heavy soils, or where there is a high level of vegetable crop residues or straw. However, the main problem occurs when the slugs or snails become mixed with the peas at harvest. Their presence leads to crop rejection by the processors or customer complaints to the retailer.

Pest Cycle
Slugs breed all through the year but most eggs are laid in the soil during the autumn. Slugs are hermaphrodite and populations tend to be aggregated in areas within a field. They feed on a wide range of plant material. Winter survival is aided by the presence of straw residue from the previous crop. Snails do not live below ground and spend the winter in uncultivated edges of the fields. They can climb foliage and stems, and migrate to crops from the field edges using overhanging plants to cross into the crop.

Prevention and Control
Straw from the previous crop should be chopped and distributed over the stubble surface, before ploughing the land in early autumn prior to pea planting the following spring. Rotations including vegetables and oilseed rape also favour slug populations. Slug populations can be monitored using a plastic dish covering a small quantity of dry poultry mash, placed on the moist soil surface within the crop. The following day, slugs can be found on the underside of the saucer and an estimate of risk made.

If high populations are found in early spring, molluscicide pellets can be applied but they will be only partly successful. The most effective timing for application is at early flower, but applications to peas which have begun to form pods should be avoided to reduce the risk of pellet contamination in harvested peas. Where slugs are seen to be actively feeding on the foliage at harvest time, mechanical harvesting should not be carried out at night or when the foliage is wet.

To reduce snail migration from overhanging or tangled vegetation from the field edges, a cultivated strip of about 2 m should be maintained between the crop and the field margin.

STEM NEMATODE:
Ditylenchus dipsaci

Host Crops
Field beans, broad beans.

Symptoms of Infestation
Damage is often first observed as plants reach flowering although young seedlings may develop severe symptoms before this stage, particularly during wet seasons. The plants are stunted and the stems thickened and twisted (**227**). There is often a blistering of the stems (**228, 229**) and this may then develop a reddish colour. Infested stems may twist and break off. Leaves may also be thickened and distorted especially at the leaf petiole, and this too may develop a red-brown discolouration. The growing points may be distorted and the upper leaves bunch together in a callous. Pods fail to fill and seeds are poorly developed or show blackening of the seed coat. Affected plants may occur singly or in large areas within the field. Both winter and spring beans are susceptible to infestation.

The nematodes are slender and transparent but are only visible under the microscope. However, as the population multiplies within the tissue, the nematodes appear as a woolly mass (**230**).

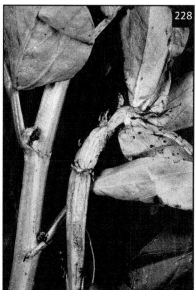

227 Distortion in a Vicia bean stem caused by stem nematode (*Ditylenchus dipsaci*).

228 Swelling in a Vicia bean stem caused by stem nematode (*D. dipsaci*) feeding.

229 Section of Vicia bean stem showing swelling and damage by stem nematode (*D. dipsaci*).

230 Stem nematodes (*D. dipsaci*) aggregating in Vicia bean stem.

Economic Importance

Stem nematode is the most damaging pest of Vicia beans in the UK and Europe. In severe cases, yield losses of up to 70% have been reported.

Pest Cycle

Stem nematodes can be both seed-borne and soil-borne. *D. dipsaci* occurs in at least two race forms in the UK, the oat race and the giant race. Although both races can affect beans, it is the giant race that causes the most severe damage. The nematodes are primarily introduced into field soil by means of infested seed. They mass under the seed coat of infested plants and desiccate as the seed matures. They survive in this state for up to 3 years but, when seed is planted, they rehydrate and invade the young tissue of the germinating bean seedling. Once in the tissue, they reproduce rapidly, producing eggs in the stem tissue. They feed within the cells on sap; the plant responds by forming giant cells which result in the stem and leaf distortion observed. The nematodes migrate within the tissue to the developing seeds in the pods. As the crop matures, infested plant material is returned to the soil where the remaining nematode population survives as free-living nematodes until the next host crop is grown. The oat race has a wide host range including bean and oat bulbs, onions, and some weed species. However, the damage caused to beans is not often severe. The giant race has a more limited host range including field and broad beans; other legume species do not seem to be affected, although they could be implicated in the survival of the pest from year to year.

Prevention and Control

The use of clean seed is a prerequisite for successfully avoiding infestations. A seed test is easily carried out and any seed lots with any level of infestation should be discarded. Once established in a crop, the seed should not be used and a break of at least 10 years is necessary before it is safe to plant beans or other host crops in infested soil again.

HELIOTHIS CATERPILLAR (CORN EARWORMS): *Helicoverpa armigera* and other *Helicoverpa* spp.

Host Crops

Green beans.

Symptoms of Infestation

The main cause of damage by the caterpillar of the Heliothis moth is by feeding on the pods of beans and boring down to the developing seed, resulting in holes in the pod surface. Such damage can lead to crop rejection. The caterpillar is initially pale green, sometimes with black dots, with a pattern of thin dark lines running along the body, the lines being darker around the second and third segments (231). The adult moth has brown forewings with a delicate darker tracery around a single dark mark on each wing. The hind wings are buff with a dark border which contains a light patch (232). In Europe, this species is often called the 'scarce bordered straw'. The moth is a frequent pest in the USA, southern Europe and Africa. Occasionally, it has arrived in large numbers in northern Europe, and crops in south west France have been damaged severely in recent years.

Economic Importance

Feeding damage can be severe in years of high pest incidence. Pod damage to beans grown for fresh picking can reduce the quality and hence the yield when damaged pods have to be removed before sale.

Pest Cycle

The pest has a very wide host range feeding on most vegetables, sweet corn and maize, and cotton bolls. The adults migrate and are often blown long distances by prevailing winds. In Europe, the main migration period is from August but has been earlier in hot summers. Eggs are laid on the foliage and these hatch in 7–10 days (**233**). The caterpillars feed for between 15 and 31 days depending on temperature and availability of food. They fall to the soil where they pupate for around 20 days. A second generation of adults may emerge and lay eggs on autumn sown crops.

Prevention and Control

Control of caterpillars with systemic insecticides is possible, although the caterpillars are difficult to control with contact acting pyrethroids. Monitoring of adults is possible using pheromone lures in funnel traps.

231 Heliothis caterpillar (*Helicoverpa armigera*).

232 Heliothis or bollworm moth (*H. armigera*).

233 First instar Heliothis caterpillar (*H. armigera*) and egg.

PEA CYST NEMATODE:
Heterodera gottingiana

Host Crops
Peas, field beans, broad beans.

Symptoms of Infestation
The pea cyst nematode, also known as pea root eelworm, is a soil-borne pest. Crops begin to show symptoms in the early part of the summer before flowering has commenced. Areas of the crop appear pale in colour and become stunted in growth. These areas are often distinct discrete patches and can range in size from a few square metres to extensive areas of the crop (**234**). Plants are often stiffly upright and small leaved. The foliage is often upwardly pointed and flowering is premature (**235**). Affected plants have poorly developed root systems with an absence of nitrogen-fixing nodules. Embedded in the roots are tiny, white, lemon-shaped cysts which later turn a chestnut coloured brown colour (**236, 237**). Infested plants become increasingly yellow and often fail to develop pods before they die prematurely. Both peas and Vicia beans can be affected by this pest.

Economic Importance
Initial damage causes yield loss of plants within infested areas. However, the spread of nematodes following an infestation results in much more widespread crop damage the next time the crop is grown in the field. Often the pest is associated with root infecting fungal infection by *Fusarium solani*. Soil-borne populations of cysts can remain viable for up to 20 years.

Pest Cycle
The cysts contain 50–100 eggs and are introduced into the field by contaminated soil on plant parts or implements. The eggs are stimulated to hatch by root exudates from the host crop (**238**). Juveniles invade the root hairs and begin feeding on cell contents of the roots. After they become mature, they mate and the male nematodes return to the soil. Eggs develop in the female's body which swells and erupts from the external root tissue, forming the lemon-shaped cysts. After the females die, the skin hardens and changes colour and in this form the cysts remain viable in the soil for many years in the absence of a host crop.

While peas are the most severely affected of the host crops of the nematode, Vicia beans can also be affected but to a much lesser degree, and only occasionally is an attack noticed in a crop. However, the nematodes reproduce readily in *Vicia faba* and when peas are grown in the field, total crop loss can occur. Other hosts include vetches (*Vicia sativa*), lupins (*Lupinus* spp.), and sweet peas (*Lathyrus oderatus*).

Prevention and Control
Nematicides can be effective if applied to known infested areas before planting, but the cost is often prohibitive. There are no commercially available varieties of peas which are resistant to pea cyst nematode. Where an infestation is noted, then a long absence of host crops is the only means of reducing the soil-borne population to a safe level. Care should be taken not to carry soil from infested fields on farm implements or plant material.

234 Damaged patch in pea crop caused by pea cyst nematodes (PCN) (*Heterodera gottingiana*).

235 Pea plant stunted and chlorotic with roots infested with PCN (*H. gottingiana*). (Courtesy of PGRO.)

236 Immature PCN (*H. gottingiana*) cysts on pea root.

237 Mature PCN (*H. gottingiana*) cysts on pea root.

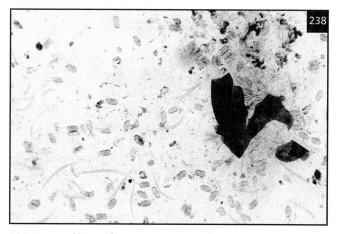

238 Eggs and larvae from an open cyst of PCN (*H. gottingiana*).

PEA THRIPS:
Kakothrips pisivorus

Host Crops
Peas.

Symptoms of Infestation
During and after flowering, heavy infestations are most likely to occur in humid conditions and where there has been a long history of pea production. Pods become silvered and distorted (**239**, **240**), and the tiny black to brown-yellow thrips can be found on pods and within the flowers (**241**). The pest is common in Europe, but related species can be found in all countries where peas are grown.

Economic Importance
Although feeding damage does not affect the yield and quality of peas grown for processing, peas grown for fresh market picking, where the appearance of the pod is important, are blemished and may be unsaleable.

Pest Cycle
The adult has an elongated body and is 1.5–2.0 mm in length. It is brown-yellow in colour, with wings usually folded along the back (**242**). Thrips are first attracted to pea flowers where they may be present in large numbers. Eggs are laid in rows on the stamen sheath and other flower parts and hatch in 7–10 days. The insects feed on the flowers and the surface of the pods, which produces the dark silvery damage symptoms. They complete their life cycle in the soil where they emerge as winged adults the following early summer, before moving to nearby peas.

Prevention and Control
In market garden situations, locally high populations are likely to build up as a result of the close proximity of the previous season's crop. Peas for processing are not significantly damaged, but crops grown for pod production should be treated when adults are found in the flowers before pods have set. Pyrethroid insecticides are useful where the thrips are exposed.

239 Pod silvering caused by pea thrips (*Kakothrips pisivorus*) feeding.

240 Pod damage caused by pea thrips (*K. pisivorus*) feeding.

241 A nymph of pea thrips (*K. pisivorus*).

242 An adult pea thrips (*K. pisivorus*).

LEAF MINERS:
Lyriomyza spp.

243 Backlit mines showing pupae and larvae of leaf miner (*Liriomyza* spp.).

Host Crops
Peas, broad beans.

Symptoms of Infestation
Leaf miners are usually active in mid- to late summer. They are widely distributed throughout the world, and several species are present in most countries. *Liriomyza sativae* is known as the vegetable leaf miner, and *L. trifolii*, the American serpentine miner. In addition, *L. congesta* and *Phytomyza horticola* are common in Europe. The most common symptoms on the leaf surface are white blisters or wide twisting tunnels where the leaf miner larvae feed and bore their way between the outer and upper epidermis of the leaf. The lower leaves are usually more heavily damaged (**243–245**).

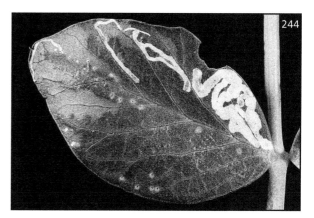

244 Leaf mines in pea leaf caused by leaf miner (*Phytomyza* spp.).

Economic Importance
In peas, the damage is usually confined to the lower leaves and while this may be unsightly, the effect on yield and quality of the peas and pods is minimal. In beans this may be similar, although heavy infestations can reduce yield of late sown crops.

Pest Cycle
Adults are small Dipterous flies. usually black in colour. Eggs are laid after the females puncture the leaf surface. The larvae hatch after a few days and are small, compressed and flattened legless larvae. The pattern of mining is characteristic to each species. When mature, the larvae pupate in golden puparia attached to the leaf and then emerge as second generation or overwinter as pupae in the soil.

Prevention and Control
There are very few occasions where chemical control is necessary in peas or beans.

245 Chrysanthemum leaf miner (*Chromatomyia syngenesiae*) mines in Phaseolus bean leaf.

ROOT KNOT NEMATODE: *Meloidogyne* spp.

Host Crops
Peas, green beans, dry beans.

Symptoms of Infestation
The growth of plants is slow and infested plants are stunted and pale, and lack vigour. In dry conditions the plants wilt and may die prematurely (**246**). Root systems exhibit swollen areas in the form of knots or galls. These may be extensive and vary in size from 1–10 mm in diameter. Severely galled roots are short and thickened and may develop into a knotted mass of distorted roots (**247, 248**). The galls should not be confused with the root nodules produced by *Rhizobium* bacteria on legume roots. These are usually pink in colour and have red central tissue.

Economic Importance
The pest is found worldwide, but frequents warm, free-draining, light soils. The effects on yield are proportional to the level of infestation; the pest is often present in soils at relatively low levels or is patchily distributed. The problems, however, can be much more severe where damage by the nematodes allows infection by *Pythium* spp.

Pest Cycle
The nematodes have a very wide host range including both crops and weeds. They survive in soil as eggs and juveniles. In moist soil conditions, newly hatched larvae and juveniles migrate towards roots and penetrate the tissue near to the root cap. Infested tissue swells as the nematodes multiply and galls eventually develop. Adult females finally develop and deposit egg sacs which survive in the soil until the next host plants are grown.

Prevention and Control
Crop rotation which includes nonhost plants such as cereals will reduce the soil-borne population of surviving nematodes. Fumigants or granular formulations of nematicides may be economic in situations where the pest is troublesome in a range of high-value crops, including beans.

246 Root knot nematode (*Meloidogyne* sp.) damage to peas.

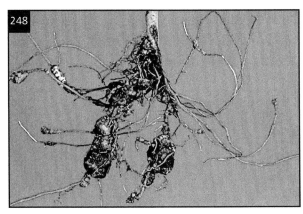

248 Root knot nematode (*M. javanica*) nodules on Phaseolus bean root. (Courtesy of George S. Awabi.)

247 Root knot nematode (*M. hapla*) galls on pea root.

TWO-SPOTTED SPIDER MITE:
Tetranychus urticae

Host Crops
Green beans, dry beans, runner beans.

Symptoms of Infestation
Damage by feeding of mites is first seen as pale patches or chlorotic spotting on the upper side of the leaves (**249**). On the underside the mites produce a fine webbing, within which mites of all growth stages can be found (**250–252**). Leaves may eventually desiccate and fall off, and bean pods may also be damaged by the feeding. The mites vary in colour from a pale pink to red, and under a lens two darker spots may be visible on the back.

249 Two-spotted spider mite (*Tetranychus urticae*) damage to Phaseolus bean leaf.

250 Two-spotted spider mite (*T. urticae*) adults on a Phaseolus bean leaf.

251 Two-spotted spider mite (*T. urticae*) infestation and webbing on a Phaseolus bean plant.

252 Two-spotted spider mite (*T. urticae*) adults on webbing.

Economic Importance

Bean yield can be reduced, as can the number of flowers, and the pod length. Blemished pods are unsaleable as fresh market beans and, when the infestation occurs early in the growth of the plants and particularly during periods of moisture stress, the plants may be severely affected.

Pest Cycle

Spider mites are widely distributed worldwide, and have an extremely wide host range which includes weeds, flowers, and fruit and vegetable crops. They are active and are able to disperse widely amongst neighbouring plants. It is possible that the mites can be blown by wind some distances from infested crops. Adults overwinter in leaf litter and other vegetable debris. In runner beans where bamboo supporting poles are used, they overwinter in large numbers in the hollow canes. Eggs are laid on leaves and juveniles go through three moults before the females lay eggs. Reproduction is both sexual and asexual.

Prevention and Control

Because of the presence of all stages of the life cycle on infested plants, chemical control is difficult and requires several applications of insecticides at intervals. Mites are also able to develop resistance very quickly to some insecticides. Biological control using predatory mites such as *Phytoseiulus persimilis* or *Amblyseius fallacis* is commonly used in both protected and outdoor crops of runner beans. Steam or gas sterilization of poles or canes is also useful for reducing the overwintering population, and the provision of regular overhead irrigation and an adequate nitrogen fertilizer helps to reduce the build-up of high populations. The integration of both cultural and biological controls is more commonly used in high-value cropping situations.

STUBBY ROOT NEMATODE: *Trichodorus* spp. and *Paratrichodorus* spp.

Host Crops

Peas, green beans, dry beans.

Symptoms of Infestation

Distinct areas and patches of crop are stunted (253, 254). The effects are more pronounced in mid-summer when crops begin flowering. Root systems may be bushy and the end of the root tips slightly swollen or distorted. Proliferation of secondary roots is also present (255). The problem is more likely to occur on sandy soils. Damage is likely following a wet spring period.

Economic Importance

The pest is widespread in North America and Europe. Where damage is extensive, then yield loss is often experienced. However, the nematodes are restricted to sandy soils; damage is sporadic because nematodes are encouraged to graze on roots when the soil is wet for long periods in the spring. In some cases, the nematodes are viruliferous and can transmit pea early browning virus to peas.

Pest Cycle

The nematodes are free-living in the soil, and migrate up and down the soil profile with the availability of soil moisture. Eggs are laid in the soil near to the roots of the host and the juveniles begin to feed at the root tip. The life cycle can be complete in 17–45 days.

Prevention and Control

Because the nematodes favour coarse sandy soils, these are the fields where damage may occur. However, the combination of nematode presence and soil moisture at the time of root development must correspond before damage occurs. In some situations, nematicides can be used as preventative treatments.

253 Stubby root nematode (*Trichodorus* sp.) field damage caused by nematode infestation.

254 Stubby root nematode (*Trichodorus* sp.) field damage caused by nematodes' root feeding.

255 Pea roots damaged by stubby root nematode (*Trichodorus* sp.)

SECTION 7

Seedling and Crop Disorders

- HOLLOW HEART
- MANGANESE DEFICIENCY
- SEED VIGOUR
- SULPHUR DEFICIENCY
- WATER CONGESTION
- IRON DEFICIENCY

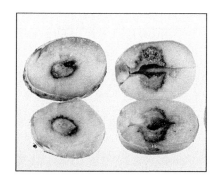

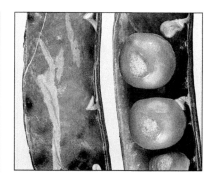

HOLLOW HEART

Affected Crops
Peas.

Symptoms of Disorder
Hollow heart, or cavitation, is a physiological disorder which results in the appearance of hollows and, in severe cases, deep fissures in the adaxial surfaces of the cotyledons of germinated pea seeds (**256**). Affected seedlings are often weak and stunted compared to normally developed seedlings. Poor seedling establishment in early sowings of peas has been associated with the presence of high levels of hollow heart affected seedlings. (See seed vigour.)

Economic Importance
Reduced seedling establishment can occur when high levels of hollow heart are present in the seed.

Prevention and Control
Hollow heart appears to be mainly associated with wrinkle seeded vining pea varieties. The disorder has been associated with those seed crops which have been subjected to high daytime temperatures during the late maturation stages of the seed.

While an examination of germinated seedlings can be made during the routine germination test, most severely affected seed lots can be detected by the electrical conductivity test used routinely to assess seed vigour.

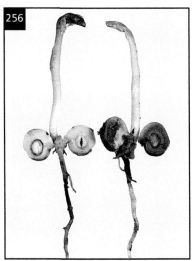

256 Hollow heart of pea. (Courtesy of PGRO.)

MANGANESE DEFICIENCY

Affected Crops
Peas, field beans, broad beans, green beans, dry beans.

Symptoms of Disorder
Manganese deficiency can develop at any time during the early stages of crop growth, and foliar symptoms can begin to show during the pre-flowering, vegetative stage. Growth appears pale in colour and closer observation of the leaves and stipules reveals a general chlorosis which is interveinal. The margins of the leaves are also chlorotic and, in time, small brown speckles develop on the leaf surface (**257**). The deficiency can show up in patches over the field or it may follow areas of compaction or soil structure differences, especially in alkaline soils. The most seriously affected crop is peas. Where plants develop manganese deficiency early on, seeds can develop an area of granulation or necrotic spot in the centre of the cotyledons (**258**). This symptom is known as marsh spot and was first described in peas grown in the Romney Marsh area of Kent in southern England. The spot may be discrete or extensive, and may affect the embryonic plumule (**259**). When marsh spot affected seeds are germinated, such damage to the plumule gives rise to abnormally developed seedlings (**257, 260**).

Economic Importance
Manganese deficiency can affect yield of peas when the effects develop early on. However, the main effect of marsh spot in the harvested produce is to reduce its value as a human consumption pea because of the blemishing (**261**). Marsh spot will also reduce the germination capacity of seed crops. In beans, yield may be reduced by a severe deficiency.

Prevention and Control

Manganese deficiency occurs when the element is locked up in alkaline soil and becomes unavailable to the plant. Where soils have a pH of 7.0 or higher, then the risk of deficiency is high. Deficiency can be corrected by foliar applications of manganese sulphate applied as soon as leaf symptoms are seen.

However, to prevent marsh spot forming in the produce, the application needs to be made during the flowering and pod development stages and repeated 14 days later. In wet summers, where prolonged growth occurs, a third application may be necessary.

257 Interveinal chlorosis caused by manganese deficiency in peas.

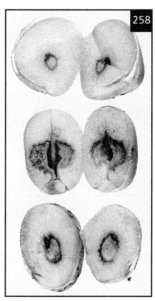

258 Marsh spot in peas caused by manganese deficiency.

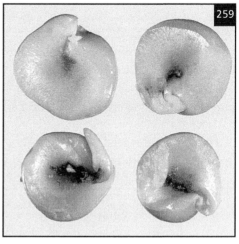

259 Seed damage due to marsh spot (manganese deficiency).

260 Multiple shooting of seedling affect by marsh spot (manganese deficiency).

261 Marsh spot (manganese deficiency) causing severe effects on peas in the pod.

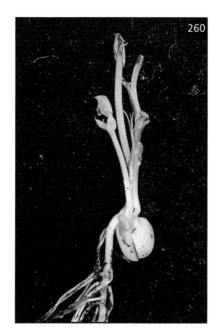

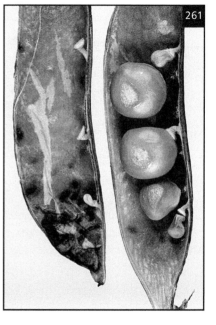

SEED VIGOUR

Affected Crops
Peas, green beans.

Symptoms of Disorder
During periods of cold wet conditions immediately after planting seed, seedling emergence is reduced, seeds rot in the soil, and plant establishment is severely affected (**262, 263**).

Imbibition, germination and seedling emergence take place over a period of up to 5 weeks. Conditions of stress, such as cold wet seedbeds, can make seeds more susceptible to attack by soil-borne pathogens, particularly *Pythium* spp.

The ability of a seed to establish into a seedling under these conditions is known as seed vigour. It is a term used to describe various aspects of the seed's ability to survive. Legume seeds are very susceptible to pre-emergence mortality, as peas and beans are often sown early in the season when conditions are not ideal. One of the factors associated with seed vigour is the integrity of the seed coat and the condition of the cells of the cotyledon. When seeds imbibe, water is absorbed firstly through the micropyle and later through the seed coat. The cells of the cotyledon rehydrate and the process of germination begins. During imbibition, however, there is a loss of carbohydrate and inorganic salt (mostly potassium) from the cotyledon to the surrounding soil water. Where this loss is high, the carbohydrate becomes a potential carbon source of soil pathogens, including *Pythium*. Where losses of sugars are high, this indicates that the cotyledon tissue is damaged; these damaged sites become colonized by *Pythium* which quickly invades the cotyledon and infects the seedling, resulting in pre-emergence decay.

Economic Importance
Wrinkle seeded vining peas and green beans are particularly susceptible to pre-emergence loss.

Prevention and Control
Seed vigour can be assessed using a laboratory seed test which measures the electrical conductivity of water which had previously had seed soaking for 24 hours. Where electrical conductivity measurement is high, the total leachates, including the sugars, are also high and this has been correlated with the ability of the seed to survive in the field. The greater the loss of leachates, the poorer is the seed vigour. This test is used throughout Europe and the USA and Australasia to choose those seedlots which are more suitable for early plantings.

262 Poor emergence caused by low pea seed vigour. (Courtesy of PGRO.)

263 Patchy emergence of low seed vigour seed in wet soil.

SULPHUR DEFICIENCY

Affected Crops
Peas, field beans.

Symptoms of Disorder
Sulphur deficiency is becoming more common in Europe where deposition of sulphur from industrial sources is reduced. Pea growth is stunted, but the most obvious effect is a pale colouration of the foliage and an upright growth of the plants (**264**). Roots may be normally developed and white. Sulphur deficiency reduces yield of peas but those grown for dry harvest may be more severely affected because of the longer growing time.

Economic Importance
Yield loss is the main effect of sulphur deficiency. While peas and Vicia beans do not respond to sulphur deficiency as much as Brassica crops (oilseed rape), the effects are similar to those of cereals.

Prevention and Control
Light, free-draining soils in areas where sulphur deposition from the atmosphere is low are more likely to be low in sulphur. Sulphate is very soluble and leaches readily from these soils. In many areas, sulphur should be considered as a major nutrient. Foliar applications of sulphur are of limited value as the sulphur is not available quickly to correct a deficiency. Pre-drilling soil applications are the most effective means of supplying sulphate. Peas require around 35–50 kg of trioxide sulphur; this is best achieved by calcium sulphate or magnesium sulphate at 110 kg/ha or by ammonium or potassium sulphate at 80–100 kg/ha. Elemental sulphur is slow to activate, although products with a very small particle size are useful.

264 Pale growth of peas caused by sulphur deficiency. (Courtesy of Rothamsted Research.)

WATER CONGESTION

Affected Crops
Peas.

Symptoms of Disorder
This is a physiological effect which causes a rupturing of the apical cells of the developing leaflets of conventionally leafed varieties of peas. The symptoms can be found about 7-10 days after a period of heavy rain in mid-summer, when the plants are growing rapidly. The leaf tips of the newest leaves are pinched and necrotic; they may at first be slightly slimy, but quickly dry to a crisp point. The next set of leaves is usually unaffected (265).

Economic Importance
The damage is very slight and only affects one set of leaves. There is no detrimental effect on the growth.

Prevention and Control
Peas growing in fertile soils are more likely to show the effects. When the humidity is high and a recent rain has fallen, the cells in the developing leaf tips are unable to transpire and rupture of the cell wall occurs. As the leaflets expand, the pinched tips are observed. There are no means of prevention and the damage is insignificant.

265 Effect of water congestion on pea foliage. (Courtesy of PGRO.)

IRON DEFICIENCY

Affected Crops
Peas.

Symptoms of Disorder
Iron deficiency is occasionally seen in crops grown on fertile, free-draining soils of high alkalinity. The main reports of deficiency have come from UK and France. The most obvious effect is a marked chlorosis of the upper foliage, which can occur at any time during the pre-flowering to pod setting growth stages (266, 267). Chlorosis can extend from the middle of the plant to the terminal shoots. On closer examination, the chlorosis can appear to be faintly mottled. Roots are normally developed and the affected areas of the crop may be associated with a more open soil structure. The effects may be seen after a period of optimum growing conditions. After about 2 weeks the effects become less obvious and the plants regain the normal green colouring.

Economic Importance
Iron deficiency is transitory and rarely results in a yield reduction.

Prevention and Control
Light, free-draining and alkaline soils may result in plants or areas of crop becoming temporarily deficient in iron. Deficiency only becomes obvious following a period of rapid growth when the availability of iron, which is a constituent of chlorophyll, becomes temporarily locked up in the alkaline soils. Iron deficiency is determined by tissue analysis but, as the effect is temporary, treatment is not worthwhile.

266 Iron deficiency causing very yellow upper part of plants in large areas of crop. (Courtesy of PGRO.)

267 Iron deficiency causing yellow upper part of pea plants. (Courtesy of PGRO.)

Further Reading

Compendium of Pea Diseases and Pests (2001). JM Kraft, FL Pfleger (eds). American Phytopathological Society, St Paul Minnesota, USA

Compendium of Bean Diseases (2005). HF Schwartz, JR Steadman, R Hall, R Forster (eds). American Phytopathological Society, St Paul Minnesota, USA

Crop Pests in the UK (1992). M Gratwick (ed). Chapman & Hall, London, UK

Crop Protection Compendium (2000). CD ROM www.cabi.org/compendia/cpc/

Diseases of Vegetable Crops (1952). JC Walker. McGraw-Hill, New York, USA

European Handbook of Plant Diseases (1988). IM Smith, RA Lelliot, DH Phillips, SA Archer (eds). Blackwell Scientific Publications, London, UK

Expanding the Production and Use of Cool Season Food Legumes: Proceedings of the Second International Food Legume Research Conference on Pea, Lentil, Faba Bean, Chickpea, and Grasspea, Cairo, Egypt, 12–16 April (1992). FJ Muehlbauer, WJ Kaiser (eds). Kluwer Academic Publishers, Dordrecht, Netherlands

The Garden Detective CD ROM (2005). (Identification & control of pests and diseases on fruit and vegetables). www.gardendetective.com British Crop Production Enterprises, Farnham, UK

Grain Legume Crops (1997). RJ Summerfield, EH Roberts (eds). Blackwell, London, UK

The Pathology of Food and Pasture Legumes (1998). DJ Allen, JM Lenné (eds). CAB International Wallingford, Oxford, UK

The Pea Crop: A Basis for Improvement (1985). PD Hebblethwaite, MC Heath, TCK Dawkins (eds). Butterworth-Heinemann, London, UK

PGRO Field Bean Growing Handbook (1994). CM Knott, AJ Biddle, BM McKeown. Processors and Growers Research Organisation, Peterborough, UK

PGRO Pea Growing Handbook (1985). AJ Biddle, CM Knott, GP Gent. Processors and Growers Research Organisation, Peterborough, UK

Plant Diseases On-line. www.inra.fr/hyp3/diseases

Plant Pests On-line. www.inra.fr/hyppz/pests

Plant Viruses On-line. The VIDE data bank for plant viruses. www.image.fs.uidaho.edu/vide/refs.htm

Recognition and Management of Dry Bean Production Problems (1983). DS Nuland, HF Schwartz, RL Forster (eds). North Central Regional Extension Publication 198, USA

A Textbook of Plant Virus Diseases (1972). KM Smith. Longman, London, UK

Vegetable Crop Pests (1992). RG McKinlay (ed). Macmillan Press, Basingstoke, UK

Index

Acanthoscelides obtectus 98
Acyrthosiphon pisum (pea aphid) 76, 78, 79, 82, 84, 86, 87, 90–1
afila types 12
Agriotes spp. 26
Agrotis segetum 92
alfalfa 38, 87
alfalfa mosaic virus 74, 86
allergies 10
Alternaria leaf spot (*A. alternata*) 36
Amblyseius fallacis 116
anthocyanins 20
anthracnose 46–7
Aphanomyces eutiches 37–8
aphids
    black bean 74, 76, 79, 93–4
    pea 76, 78, 79, 82, 84, 86, 87, 90
    peach potato 74, 76, 79, 82, 84

Aphis fabae (black bean aphid) 74, 76, 79, 93–4
Apion vorax 75
Ascochyta bolthauseri 61–2
Ascochyta fabae 38–9
Ascochyta leaf spot 61–2
Ascochyta phaseolorum 61–2
Ascochyta pisi 40–2
Autographa gamma 95–6
azoxystrobin 42, 69

bacterial blight, pea 64–5
bean common mosaic virus 76
bean curly top virus 77
bean leaf and pod spot 38–9
bean leaf roll virus 78, 87
bean seed beetle 98
bean seed fly 27–8
bean yellow mosaic virus 79
beetles
    bean seed 98
    click 26
    pea seed 96–7
biological control methods 116
black root rot 23–4
blight, common 72
bollworm 108–9
Botrytis cinerea 42–4, 46, 90
Botrytis fabae 44–6
brassica crops 32, 123
broad bean stain virus 75
broad bean true mosaic virus 75
Bruchus pisorum 96–7
Bruchus rufimanus 98

canning 12, 13

caterpillars
    Heliothis moth 108–9
    pea moth 103–4
    silver Y moth 95–6
    tortrix moth 100
    turnip cutworm 92
cavitation (hollow heart) 120
Cepaea sp. 105–6
Cercospora spp. 44
Cernuella sp. 105–6
chlorothalonil 42
chocolate spot 44–6, 70
Chromatomyia syngenesiae 113
Circulifer tenellus 77
click beetle 26
clover seed weevil 75
clovers 38
Cnephasia asseclana 100
Colletotrichum lindemuthianum 46–7
combining peas 12
Contarinia pisi 101–2
copper fungicides 63
crane fly 32–3
crop protection 14
crop rotation 14
    brassicas 32, 123
    Fusarium wilt control 50
    nonlegumes 42, 53
crops
    breeding 12–13
    origins and forms 10–11
    production 12–14
cucumber mosaic virus 80
cutworms 92
Cydia nigricana 103–4
cymoxanil 20, 58, 60
cyproconazole 42

damping off 20
deficiencies
    iron 124
    manganese 120–1
    sulphur 14, 123
Delia platura 27–8
Deroceras reticulatum 105–6
diagnosis, quick guide 16–18
Didymella fabae 38–9
disease resistance/tolerance
    bacterial blight 64
    Fusarium wilt 50
    pea streak virus 86
    powdery mildew 48
dithiocarbamates 58
Ditylenchus dipsaci 107–8

downy mildew
    bean 57–8
    pea 59–60

Erysiphe pisi 48–9
Euonymus europaeus 94

Faure, Madame 12
'favism' 10
field viners 12
fludioxonil 42, 53
food value 10–11
foot rot
    bean 55–6
    pea 52–3
forecasting systems 14, 90
fosetyl aluminium 58, 60
frost 64
frozen peas 12
Fusarium culmorum 55–6
Fusarium oxysporum f.sp. *pisi* 50
Fusarium root rot (Fusarium yellows) 54
Fusarium solani 55–6, 110
Fusarium solani f.sp. *phaseoli* 54
Fusarium solani f.sp. *pisi* 52–3
Fusarium wilt 50–1

grasslands, cultivation 26, 32–3
grey mould 42–4, 90

halo blight 62–3
harvesting 12, 13
Helicoverpa armigera 108–9
herbicides 46
Heterodera gottingiana 110
hollow heart 120
hoverflies 90
humidity 44, 124

iron deficiency 124
irrigation 12, 44, 62, 64

Kakothrips pisivorus 112

lambda-cyhalothrin 102
leaf hopper, beet 77
leaf miners 113
leaf notching 28, 29
leaf and pod spot
    bean 38–9
    pea 40–2
leaf spot
    Alternaria 36
    Ascochyta 61–2

leatherjackets 32–3
leaves, vein blackening 46, 47
lima bean 11
Lyriomyza spp. 113

Macrosiphum euphorbiae 76
manganese deficiency/marsh spot
  120–1
Meloidogyne spp. 114
metalaxyl 20, 58, 60
midges
  pea gall (*Contarinia pisi*) 101–2
  Reseliella sp 56
mildew *see* downy mildew; powdery
  mildew
mites
  predatory 116
  two-spotted spider 115–16
monitoring systems 14, 30, 102, 104,
  109
moths
  Heliothis/bollworm 108–9
  pea 103–4
  silver Y 95–6
  tortrix 100
  turnip cutworm 92
mould *see* grey mould; white mould
Mycosphaerella pinodes 40–2
Myzus persicae (peach potato aphid)
  74, 76, 79, 82, 84

nanism 30
nematodes
  pea cyst 110
  root knot 114
  stem 107–8
  stubby root 81, 82, 116
  transmission of viruses 81, 82, 116
nitrogen fixation 13
nutritional requirements, crop 14

Paratrichodorus spp. 116
pastures *see* grasslands
pea and bean weevil 28–30, 75, 79
pea dwarfing syndrome 30
pea early browning virus 81–2, 116
pea enation mosaic virus 82
pea foot rot 52–3
pea leaf and pod spot 40–2
pea leaf roll virus 87
pea seed beetle 96–7
pea seed-borne mosaic virus 84
pea streak virus 86
pea top yellows virus 87
Peronospora viciae 57–60
pest monitoring 14, 30, 102, 104, 109
pesticides 14
  fungicides 20, 39, 42, 44, 46, 48,
    58, 60, 63, 69, 70
  insecticides 28, 32, 92, 96, 97, 98,
    102, 112

pesticides (*continued*)
  molluscide pellets 106
  resistance to 42, 44, 60
Phaseolus vulgaris (lima bean) 11
Phaseolus beans 13
Phaseolus coccineus (runner bean) 11,
  13, 28, 61, 116
Phaseolus lunatus 11
phenols 20
phenylanylines 58
pheromone traps 30, 102, 104, 109
Phoma exigua var. *exigua* 61
Phoma medicaginis 40, 41, 42
Phoma medicaginis var. *pinodella*
  52–3, 55–6
Phytomyza spp. 113
Phytoseiulus persimilis 116
pirimicarb 90
planting 13
plough pan 37, 38
pod rot, Botrytis 42–4
pod spot
  Alternaria alternata 36
  Ascochyta spp. 38–9, 40
  Mycosphaerella pinodes 40–2
  Xanthomonas campestris 72
powdery mildew 48–9
Pseudomonas syringae pv.
  phaseolicola 62–3
Pseudomonas syringae pv. *pisi* 64–5
pyrethroid insecticides 92, 96, 112
Pythagoras 10
Pythium spp. 122
  root rot 22
  seedling rot/damping off 20
Pythium ultimum 20–1

red clover vein virus 86
Reseliella sp. 56
Rhizobium bacteria 13, 28, 29, 114
Rhizoctonia spp. 22
root galls, nematodes 114
root nodules (*Rhizobium*) 13, 114
  pest damage 28, 29
root rot
  black 23–4
  common 37–8
  Fusarium spp. 54
  Pythium spp. 22
  Rhizoctonia spp. 22
runner bean 11, 13, 28, 61, 116
rust, bean 69–70

Sclerotinia sclerotiorum 66
Sclerotinia trifoliorum 68
seed corn maggot 27–8
seed germination 14, 122
seed testing 42, 53, 122
seed treatments
  downy mildew 58, 60
  halo blight 63

seed treatments (*continued*)
  pea leaf and pod spot 42
  Phoma infection 53
  powdery mildew 58
seed vigour 122
Sitona lineatus 28–30, 75, 79
slugs and snails 105–6
soils 14, 37, 38
  alkaline 14, 24, 121
  disease testing 53
  nutrient deficiencies 14, 121, 123
spider mite, two-spotted 115–16
spindle, European 94
stem rot
  Pythium spp. 20
  Rhizoctonia 22
  Sclerotinia 68
Stemphylium spp. 44
streptomycin sulphate 63
strobilurins 70
sulphur deficiency 14, 123
sulphur treatments 48
symptoms of disease 16–18

tebuconazole 69
Tetranychus urticae 115–16
Thanatephorus cucumeris 22
thiabendazole 42, 53
Thielaviopsis basicola 23–4
thiram 20
thrips
  field 30–2
  pea 112
Tipula spp. 32–3
Triaspis luteipes 98
triazole-based fungicides 48
triazoles 70
Trichodorus spp. 81, 82, 116

Uromyces appendiculatus 69
Uromyces fabae 70

Vicia faba 10, 13
vinclozolin 42
vining peas 12

water congestion 124
weeds 80, 82, 87, 94, 110
weevil
  clover seed 75
  pea and bean 28–30, 75, 79
white mould 66
white-flowered varieties 20
wild hosts 48
wilt, *Fusarium* 50
wireworms 26

Xanthomonas campestris pv. *phaseoli*
  72

Printed and bound by CPI Group (UK) Ltd, Croydon, CR0 4YY

23/10/2024

01778247-0009